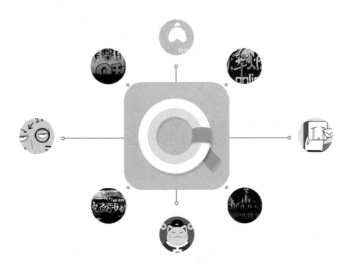

Getting Started with CrossApp

CrossApp
入门指南

沈大海　张 磊◎编著
Shen Dahai　Zhang Lei

清华大学出版社
北京

内 容 简 介

本书系统论述了CrossApp跨平台App开源引擎的开发理论与实践。全书内容涵盖了CrossApp引擎特点,开发环境设置,核心类CAVeiw、CAViewController和CAWindow的实现原理和使用,CrossApp内存管理机制,CrossApp核心控件使用,设备功能调用及网络通信功能等。本书共8章,分为如下三大部分。

第一部分为开发基础,即第1~4章,内容包括CrossApp开发环境搭建,引擎原理介绍,创建项目和核心UI组件类的使用。通过该部分内容的学习,读者可以创建一个简单的跨平台应用。

第二部分为开发进阶,即第5~7章,内容包括在CrossApp项目中使用多媒体功能,使用文件存储功能,实现网络功能,实现UI组件的动画效果。通过该部分内容的学习,读者可以实现一款功能强大的跨平台应用。

第三部分为项目实战,即第8章,通过一款手机电商App的源码解析,可以让读者具备架构一款大型跨平台联网App的能力。

本书封面贴有清华大学出版社防伪标签,无标签者不得销售。
版权所有,侵权必究。侵权举报电话:010-62782989 13701121933

图书在版编目(CIP)数据

CrossApp入门指南/沈大海,张磊编著.--北京:清华大学出版社,2016
(清华开发者书库)
ISBN 978-7-302-42134-4

Ⅰ.①C… Ⅱ.①沈… ②张… Ⅲ.①移动终端-应用程序-程序设计 Ⅳ.①TN929.53

中国版本图书馆CIP数据核字(2015)第267383号

责任编辑:盛东亮
封面设计:李召霞
责任校对:胡伟民
责任印制:杨 艳

出版发行:清华大学出版社
    网 址:http://www.tup.com.cn, http://www.wqbook.com
    地 址:北京清华大学学研大厦A座 邮 编:100084
    社 总 机:010-62770175 邮 购:010-62786544
    投稿与读者服务:010-62776969,c-service@tup.tsinghua.edu.cn
    质量反馈:010-62772015,zhiliang@tup.tsinghua.edu.cn
    课件下载:http://www.tup.com.cn,010-62795954
印 装 者:北京嘉实印刷有限公司
经 销:全国新华书店
开 本:186mm×240mm 印 张:9.75 字 数:247千字
版 次:2016年1月第1版 印 次:2016年1月第1次印刷
印 数:1~2000
定 价:39.00元

产品编号:065985-01

序
FOREWORD

  CrossApp 的读者大家好，在得知这本 CrossApp 图书即将出版的消息后，我兴奋不已。

  CrossApp 作为 9 秒社团推出的一款开源跨平台应用开发引擎有其显著的优势：速度快，更安全以及功能强大，越来越多的 App 开发者在关注该引擎的发展，也有一部分移动互联网创业团队在使用 CrossApp 来创建可以在 Android 和 iOS 平台上运行的 App 产品（当前 CrossApp 还支持更多的平台）。

  这套图书系统介绍了 CrossApp 的开发技术，并详细讲解了我们推出的电商开源项目，读者可以通过本书快速掌握开发 CrossApp 应用技巧。非常感谢沈大海老师和相关的开发人员为 CrossApp 的推广做出的贡献，也希望越来越多的开发者加入 CrossApp 大家庭。

  9 秒社团的官方网址为 www.9miao.com。

李　翀

前言
PREFACE

移动互联网发展迅速，截至 2015 年 1 月，已经有超过 120 万款应用上线苹果 AppStore，这其中包含了游戏、电商、社交和工具等类型的产品，越来越多的应用在 iOS 平台运营成功之后都会希望移植到 Android 系统以及其他移动终端系统，这为开发者带来了巨大的时间成本和资金消耗。

CrossApp 通过跨平台的解决方案，可以通过 C++ 和 JS 等语言实现一处开发及多处发布的功能，并且在功能和性能上有明显优势，为开发移动应用的项目提供了全新的解决方案，这对于在移动互联网创业的中小团队来讲，无疑是雪中送炭。

CorssApp 是一套跨平台的开源技术引擎，源码完全开放，可以免费使用。引擎中提供了开发移动 App 所需要的 UI 框架、底层设备访问功能、网络通信框架和组件动画功能，这对于有一定移动开发基础的程序员来讲可以极大提升开发效率。

本书首先介绍 CrossApp 开发环境搭建、引擎原理、项目创建和核心 UI 组件类的使用；之后介绍多媒体功能、文件存储功能、网络功能和 UI 组件的动画效果等高级用法；最后通过在 www.9miao.com 发布的手机电商 CrossApp 的源码解析，让读者具备架构一款大型跨平台联网 App 的能力。

本书编写过程中得到了 CrossApp 引擎的主程栗元锋的大力支持，特此感谢。

通过本书的学习，希望读者了解 CrossApp 跨平台引擎的特点以及能够使用该引擎搭建一款跨平台 App 的产品架构，由于编写仓促，书中难免有疏漏与不妥之处，敬请读者批评指正。如果读者在学习过程中有任何问题可以发送邮件到 shendahai@longtugame.com 或 zhanglei1@longtugame.com，我们会及时回复，读者也可以登录 edu.9miao.com 来同步学习相关课程的视频。

作　者
2016 年 1 月于北京

目 录
CONTENTS

第1章 CrossApp 简介及开发环境搭建 1
- 1.1 CrossApp 简介 1
  - 1.1.1 CrossApp 功能与特色 1
  - 1.1.2 CrossApp 的优势 2
- 1.2 CrossApp 开发环境搭建 3
  - 1.2.1 Windows 开发环境搭建 3
  - 1.2.2 在 Windows 系统创建 CrossApp 工程 5
  - 1.2.3 CrossApp 项目目录说明 6
  - 1.2.4 通过 Visual Studio 2013 启动一个项目 7
  - 1.2.5 Windows 环境下 Android 配置 9
  - 1.2.6 Mac OS X 开发环境搭建 12
  - 1.2.7 Mac 下配置 Android 开发环境 13
  - 1.2.8 第一个 CrossApp 项目解析 14

第2章 CrossApp 基础概念 19
- 2.1 核心类 19
  - 2.1.1 CAView 19
  - 2.1.2 CAViewController 21
  - 2.1.3 CAWindow 22
- 2.2 内存管理 22
  - 2.2.1 对象内存引用记数 22
  - 2.2.2 手工对象内存管理 24
  - 2.2.3 自动对象内存管理 24
- 2.3 坐标系 26
- 2.4 适配方案 29
- 2.5 深入理解 CAViewController 和 MVC 30
  - 2.5.1 CAViewController 的职责 30

2.6　CAViewController 类的使用 ･･････････････････････････････････････ 32
　　2.6.1　CAViewController 生命周期 ･･･････････････････････････････ 33
　　2.6.2　CAViewController 使用 ･･･････････････････････････････････ 33
2.7　CANavigationController 导航视图控制器 ････････････････････････ 37
2.8　CATabBarController 切换视图控制器 ･･･････････････････････････ 39
2.9　CADrawerController 侧边抽屉式导航控制器 ････････････････････ 42

第 3 章　CrossApp 核心控件与视图 ･･･････････････････････････････････ 44

3.1　文本 CALabel ･･ 44
3.2　按钮 CAButton ･･ 46
3.3　图片 CAImageView ･･ 52
3.4　九宫格图片 CAScale9ImageView ････････････････････････････････ 54
3.5　单行输入框 CATextField ･･･････････････････････････････････････ 56
3.6　多行输入框 CATextView ･･･････････････････････････････････････ 60
3.7　开关 CASwitch ･･ 63
3.8　提示框 CAAlertView ･･･ 65
3.9　进度条 CAProgress ･･･ 68
3.10　滚动条 CASlider ･･ 70
3.11　步进控件 CAStepper ･･ 72
3.12　滚动视图 CAScrollView ･･･････････････････････････････････････ 75
3.13　列表视图 CAListView ･･･ 79
3.14　表格视图 CATableView ･･･････････････････････････････････････ 86
3.15　容器 CACollectionView ･･･････････････････････････････････････ 96
3.16　切换页面 CAPageView ･･ 102

第 4 章　CrossApp 数据存储与解析 ･･･････････････････････････････････ 105

4.1　CAUserDefault 简单存储 ･･･････････････････････････････････････ 105
4.2　SQLite 的使用 ･･･ 107
4.3　JSON 解析 ･･ 109
4.4　XML 解析 ･･ 111

第 5 章　CrossApp 设备功能调用 ･････････････････････････････････････ 115

5.1　相机 ･･ 117
5.2　相册 ･･ 119
5.3　通讯录 ･･ 121
5.4　本章小结 ･･ 127

第 6 章　CrossApp 多媒体 ……128
6.1　CAViewAnimation 动画 ……128
6.2　SimpleAudioEngine 音效 ……130

第 7 章　CrossApp 网络通信 ……133
7.1　HTTP 基础使用 ……133
7.2　HTTP 加载网络图片 ……135

第 8 章　CrossApp 项目实战 ……137
8.1　折 800 开源项目介绍 ……137
8.2　项目架构设计 ……138
8.3　核心模块说明 ……140
8.4　本章小结 ……146

第 1 章 CrossApp 简介及开发环境搭建

1.1 CrossApp 简介

CrossApp 是一款完全开源，免费和跨平台的移动应用开发引擎，基于最宽松的 MIT 开源协议，开发者可以根据自身情况使用 CrossApp 开发任何商业项目。CrossApp 以 C++ 作为开发语言，图形渲染基于 OpenGL ES 2.0，采用 MVC 框架模式。使用 CrossApp 开发的应用程序支持导出到各大主流移动平台，真正实现一次编码，多处运行的跨平台开发技术。

CrossApp 主要由 9 秒社团 (www.9miao.com) 自研，官方制定了 CrossApp 的基本架构，确立了 CrossApp 的基本雏形，其后的版本也将由来自各方的开发精英自愿组成的 9 秒社团常务贡献委员会进行更新和维护。

CrossApp 的官方网站为 http://crossapp.9miao.com/。

最新版本源码的下载地址：Oschina：http://git.oschina.net/9miao/CrossApp 和 GitHub：https://github.com/9miao/CrossApp。

1.1.1 CrossApp 功能与特色

1）跨平台

如图 1-1 所示，CrossApp 主要支持目前最主流的移动平台 iOS 和 Android，后续更新版本将完善更多平台的支持。

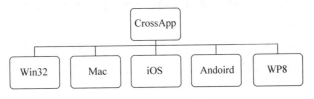

图 1-1 CrossApp 所支持的平台

2）整合第三方库

CrossApp 整合一些第三方库，例如常见的数据解析库 JsonCpp、TinyXML 和 HTTP 等。

3）基于 OpenGL ES 2.0

CrossApp 的图形渲染使用的是 Open GL ES 2.0，渲染效率高，可以使移动设备的 GPU 发挥到最佳效果。

4）丰富的 UI 控件

CrossApp 的设计宗旨在于为移动应用开发者提供快速、高效的开发解决方案。在此基础上，CrossApp 封装了大量的 UI 控件，各类控件的功能十分丰富，开发者可以直接使用这些控件进行应用的开发。这些控件基本满足应用开发需求中的大部分功能，也免去了封装 UI 控件花费的大量时间，能进一步提升开发的效率，节省开发的成本。同时 CrossApp 也可以整合部分由第三方开发者提供的优秀控件资源，更大地丰富 CrossApp 的 UI 控件。

5）CrossApp 耗电优化

由于之前 CrossApp 抽取了 Cocos2d-x 的渲染驱动模式，在程序生命周期中不断重绘，渲染驱动模式的缺点就是 CPU 占用高，因此耗电量大。这种不断的重绘方式对于游戏可能影响不大，但对于 App 来说太浪费了。因此 CrossApp 针对应用程序的特性，将渲染机制改为事件驱动模式，这种模式的渲染要有外界触发才会重绘，在没有外界触发的时候画面静止，渲染停止，以达到节能的效果。

1.1.2 CrossApp 的优势

通过表 1-1 分析可以得出如下结论。

1）Mobile Web

以 HTML5 和 JQuery 为代表的 Web 开发技术是以纯浏览器为基础的，所以没有离线能力，更无法充分发挥各平台的功能和特性。

2）Hybrid App

Hybrid App 即混合模式应用程序，是介于 Web 和 Native 之间的 App，具备一定 Native 原生 App 的优秀的用户体验和 Web App 跨平台的优势。但由于 Hybrid 仍旧以前端技术为基础，无法对内存和系统资源进行有效的管理。

3）Interpreted App

以前端技术为基础，同样不能有效地对内存和系统资源进行管理。

4）Native App

原生开发以 iOS 和 Android 为代表，虽然原生开发能够实现最佳的用户体验和高优化，但开发的成本较高。而且因为原生开发几乎不支持跨平台特性，相对于跨平台技术来说，实际效益更低。

综上所述，Native App 和 Cross Compiling 的综合效率更高，开发者需要根据实际情况，考虑各方面因素来选择合适的开发解决方案。

表 1-1 Cross App 与其他跨平台引擎对比

技术 说明	Mobile Web	Hybrid App	Interpreted App	Cross Compiling	Native App
代表产品	HTML5/JQuery Mobile	PhoneGap	Titanium	CrossApp	iOS/Android
跨平台性能	强	强	中	强	低
离线能力	无	有	有	有	有
入门要求	低	低	低	低	高
功能	弱	中	中	强	强

1.2 CrossApp 开发环境搭建

CrossApp 可以分别在 Windows 平台和 Mac OS 平台上配置开发环境，展示良好的跨平台性，我们推荐大家使用 Mac OS 平台做 CrossApp 的开发工作，这样能更方便地在 iOS 平台和 Android 平台进行测试，得到及时的 bug 反馈。

当然也可以先在 Windows 平台开发后，再转到 Mac OS 发布 iOS 版本，但需要注意转码问题。

1.2.1 Windows 开发环境搭建

Windows 环境开发 CrossApp 系统需求如下。

（1）操作系统：Windows 7 以上版本。
（2）开发工具：Visual Studio 2012 以上版本（推荐 Visual Studio 2013）。
（3）Python：推荐 Python 2.7 版本。

首先下载 CrossApp 引擎源码：

github 地址：https://github.com/9miao/CrossApp
oschina 地址：http://git.oschina.net/9miao/CrossApp

1. Python 脚本创建

1）双击安装 Python
2）配置环境 Path

右击"计算机（我的电脑）"→"属性"→"高级系统设置"→"环境变量"→"系统变量"，双击 Path，在变量值结尾添加"；C:/Python27"（引号内容为 Python 的安装路径，注意前面有一个分号）如图 1-2 所示。

3）测试 Python 环境

如图 1-3 所示在命令窗口输入 python -version，如果出现 Python 2.7.3 说明 Python 安装成功。

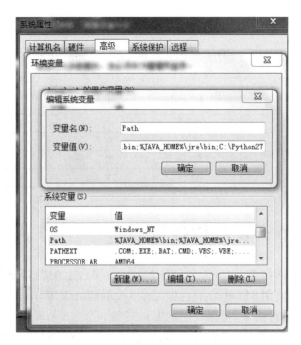

图 1-2 设置 Python 环境变量

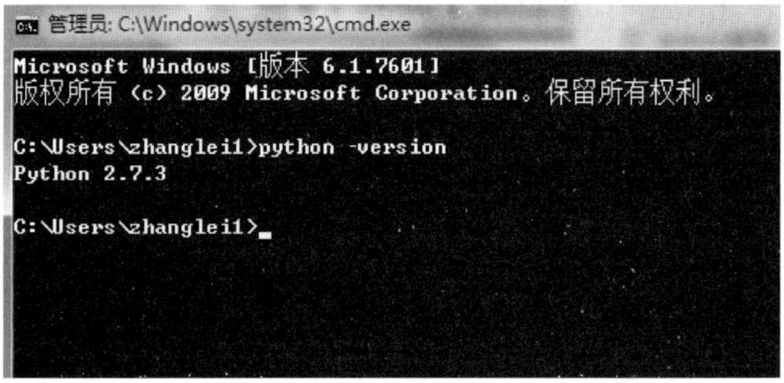

图 1-3 Python 环境配置

2．解压缩 CrossApp 项目源码

将下载的 CrossApp 项目源码解压缩，目录结果如图 1-4 所示。

（1）CocosDenshion：包含了 CrossApp 多媒体 API 的源码。

（2）CrossApp：包含了 CrossApp 引擎的核心源码。

（3）document：引擎相关文档。

（4）extensions：引擎扩展库 JSON、网络和 SQLite 等的源码。

（5）licenses：引擎的相关开源协议说明。

（6）project-creator.app：在 Mac 平台创建项目的应用程序。

图 1-4 CrossApp 项目源码目录

（7）projects：项目文件夹（该文件夹会在创建第一个项目后自动创建，源码包里并不包含）。

（8）samples：CrossApp 官方 Demo。

（9）scripting：对 JS 脚本的支持源码。

（10）template：CrossApp 项目模板。

（11）ThreePartSDK：第三方库 ShareSDK 和 UMSDK。

（12）tools：引擎相关工具

1.2.2 在 Windows 系统创建 CrossApp 工程

在 Windows 系统创建 CrossApp 工程有如下两种方法。

1. 使用工具一键创建（无须 Python 环境）

在 CrossApp 的引擎目录下，双击 project-creator.exe，弹出如图 1-5 所示窗口。

第一栏填写工程名，第二栏填写 Android 包名，然后单击 Create Project Now，完成 CrossApp 项目的创建。

创建工程会在引擎的根目录多出一个名为"projects"的文件夹，创建的工程也就在这个目录下。

图 1-5　创建第一个 CrossApp 项目

2．Python 脚本创建

1）配置环境 Path

右击"计算器(我的电脑)"→"属性"→"高级系统设置"→"环境变量"→"系统变量",然后双击 Path,在变量值结尾添加:";引擎根目录\tools\"。

2）创建工程

打开 CMD 命令提示符,输入:python create_project.py -project HelloCrossApp -package com.9miao.crossapp -language cpp,如图 1-6 所示。

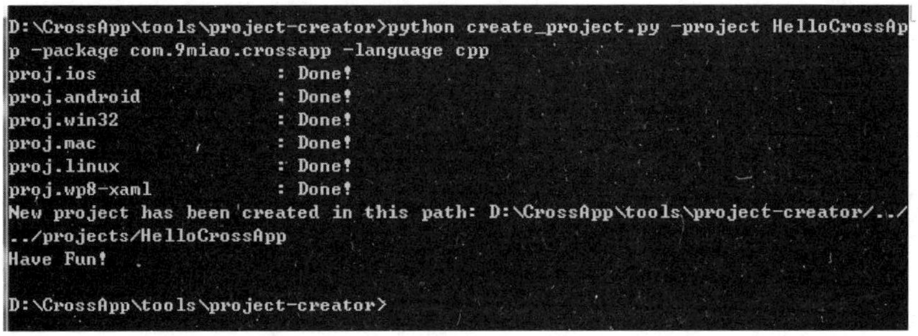

图 1-6　使用命令行创建 CrossApp 项目

看到提示成功的信息,就说明工程创建成功了。

其中命令行结构如下:

Python create_project.py -project 项目名称 -package 包名 -language 开发语言。

1.2.3　CrossApp 项目目录说明

新建完项目之后,projects 的文件夹目录结构如下:

（1）Classes：包含新项目的源码。

（2）Resources：新项目的所有资源文件,包括图片、声音和文本等。

（3）Proj.android：新项目的 Android 工程，可以使用 ADT 打开。
（4）Proj.ios：新项目的 iOS 工程，可在 Mac 平台使用 Xcode 打开。
（5）Proj.mac：新项目的 Mac 工程，可在 Mac 平台使用 Xcode 打开。
（6）Proj.win32：新项目的 Win32 工程，可在 Windows 平台使用。

1.2.4　通过 Visual Studio 2013 启动一个项目

打开项目目录下：\proj.win32\HelloCrossApp.sln（创建的工程名）。

（1）启动 Visual Studio 2013 后，右击 HelloCrossApp，设置为启动项目，如图 1-7 所示。

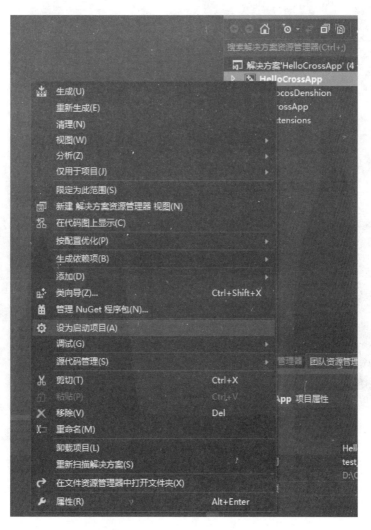

图 1-7　在 Visual Studio 2013 中打开项目

（2）右击 HelloCrossApp→"调试"→"启动新实例"，如图 1-8 所示。

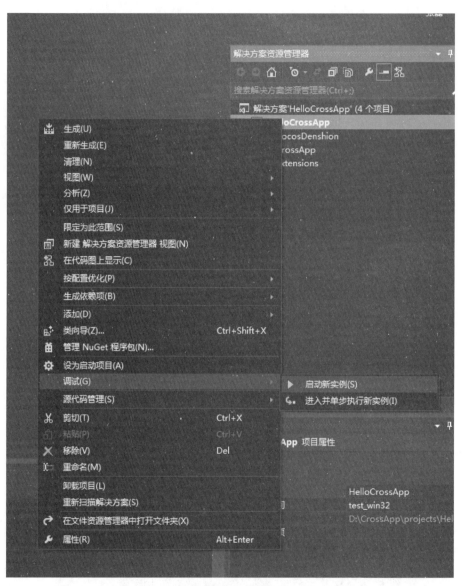

图 1-8　启动新实例

（3）第一次启动等待时间较长，如果没有错误信息，将会弹出 CrossApp 项目窗口，如图 1-9 所示。

本书以在 Windows 环境下使用 Visual Studio 2013 为开发环境介绍 CrossApp 的开发技术。

图 1-9　第一个 CrossApp 项目

1.2.5　Windows 环境下 Android 配置

移植 Android 的环境配置所需的工具如下：

（1）adt-bundle-windows-x86_64-20130917

下载地址：http://developer.android.com/sdk/index.html

（2）android-ndk-r9c

下载地址：http://developer.android.com/tools/sdk/ndk/index.html

（3）jdk-7u45-windows-x64

下载地址：http://www.oracle.com/technetwork/java/javase/downloads/index.html

以上列出的工具也可以使用其他的版本，NDK 的 r8 以上版本，配置方法基本一样，请根据操作系统型号选择配置环境所需的工具包。

1. 安装 Java 环境

配置 Android 环境必须要先安装 Java 环境,安装过程并没有特别要求。

安装完成后需要配置环境变量,在 Windows 系统变量中新建一个变量名为 JAVA_HOME 的变量,变量值为 JDK 的安装目录,例如:C:\Program Files\Java\jdk1.7.0_45。然后新建一个名为 CLASSPATH 的变量,变量的值为.;%JAVA_HOME%\lib;%JAVA_HOME%\lib\tools.jar(最前面的".;"一定要加上),在系统变量中找到 Path 变量,点击编辑,在变量值最前面添加一个分号,然后在分号的前面添加%JAVA_HOME%\bin;%JAVA_HOME%\jre\bin。这样 Java 的环境变量就配置好了。

打开 DOS,输入 java、javac 和 java -version 等命令就可以看见相关的信息。

2. 更新 ADT 以及配置 NDK

将下载好的 ADT 和 NDK 解压在指定目录,运行 ADT 目录下的 eclipse。如果提示要求更新,在 eclipse 的菜单栏中单击 Windows,选择 Android SDK Manager 在线更新 SDK,有更新的话尽量全选,更新的过程会花费较多的时间,直至 SDK 更新完成(后面的在 Xcode 下配置 Android 环境,如果有需要,同样需要经过这一步)。

3. 移植 Android

将工程导入 eclipse,File→New→Other→Android Project from Existing Code,单击 Browse 找到我们创建的工程目录下的 proj.android,如图 1-10~图 1-12 所示。

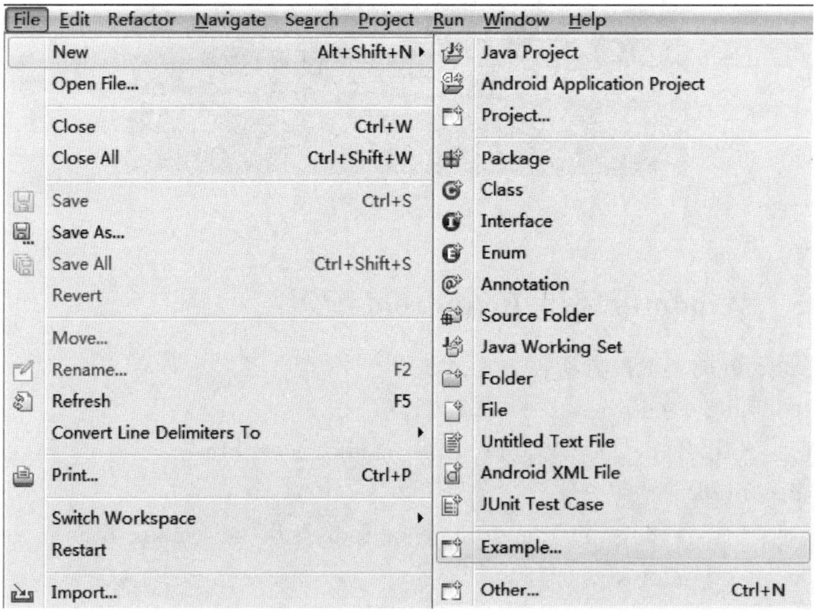

图 1-10 在 ADT 中导入 CrossApp 项目(一)

工程导入 eclipse 后,编译运行过程中所出现的异常和错误以及对应的解决办法如表 1-2 所示。

第1章　CrossApp简介及开发环境搭建　11

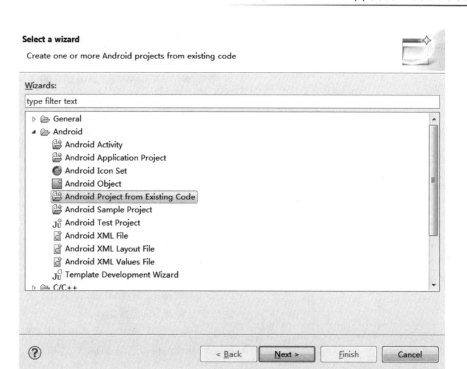

图 1-11　在 ADT 中导入 CrossApp 项目（二）

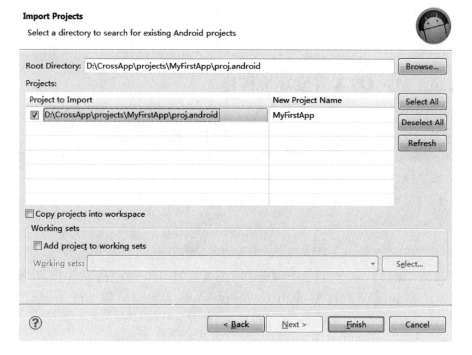

图 1-12　在 ADT 中导入 CrossApp 项目（三）

表 1-2 导入 ADT 错误排查表

错　　误	解　决　办　法
在工程 src 下的包错误，提示找不到 org. CrossApp.lib 这个包	将 CrossApp\CrossApp\platform\android\java\src 路径下的文件夹复制到工程的\proj.android\src 目录下。注意：org.CrossApp.lib 包需要和工程的包名在 src 的同级目录下，不能有包含关系
运行工程出现 Error：Program "bash" is not found in PATH	右击所建工程的 Properties，在打开面板中单击 C/C++ Build，在 Build command 一项中，把默认的 command 替换为自己 NDK 目录下的 ndk-build.cmd，例如：D:\android-ndk-r9c\ndk-build.cmd
运行工程出现 Cannot find module with tag 'CrossApp' in import path	在 android.mk 中找到 $(call import-module, CrossApp)，在这句前面添加如下两段代码： $(call import-add-path, D:/CrossApp) $(call import-add-path, D:/CrossApp/CrossApp/platform/third_party/android/prebuilt) 路径修改为对应的安装路径
模拟器运行崩溃	Android 模拟器从 SDK 4.0.3 开始才支持 OpenGL ES2.0，新建的模拟器的 SDK 也要求 4.0.3 及以上版本

4. 打包运行 Android 项目

导入项目成功后，将 Android 手机连接到开发电脑，并开启调试模式，就可以通过在项目上右击选择

run As→android Application，在 Android 手机上运行该项目。

1.2.6　Mac OS X 开发环境搭建

在 Mac OS 中使用 Xcode 开发 CrossApp 更加的方便快捷，XCode 可以在 AppStore 免费下载，安装方法也很简单，读者可以参考相关资料，此处不再赘述。

在 Mac OS 中使用 Python 来创建工程，其步骤和 Windows 下使用 Python 创建工程类似，只是一般 Mac OS 已经安装了 Python 的环境，无须再进行配置。

（1）打开"终端"，移动到 CrossApp\tools\project-creator 目录下。

（2）输入如下命令：

python create_project.py – project HelloCrossApp – package com.9miao.crossapp – language cpp

这样就在引擎的 projects 目录下创建了一个名为 HelloCrossApp 的工程，开发时请打开项目目录下的 proj.ios 文件夹，双击 HelloCrossApp.xcodeproj 即可在 Xcode 中打开新建的项目。

1.2.7 Mac 下配置 Android 开发环境

Mac OS 下的 Android 环境配置相对简单，因为本身集成了 Java 和 Python 环境，所以免去了 Java 和 Python 的配置。

1. 下载 NDK 和 ADT

下载较新版本的 NDK 和最新版的 ADT(for Mac)，解压到指定目录。

NDK 下载地址：http://developer.android.com/tools/sdk/ndk/index.html

ADT 下载地址：http://developer.android.com/sdk/index.html

2. 配置 .bash_profile 文件

打开终端输入 pico .bash_profile，打开该文件的文本编辑，在里面输入如下文本：

```
export CROSSAPP_ROOT = /Users/ * * * /Documents/CrossApp
export ANDROID_SDK_ROOT = /Users/ * * * /Documents/android/adt-bundle-mac/sdk
export ANDROID_NDK_ROOT = /Users/ * * * /Documents/android/android-ndk-r8b
export NDK_ROOT = /Users/ * * * /Documents/android/android-ndk-r8b
export PATH = $ PATH: $ ANDROID_NDK_ROOT
export PATH = $ PATH: $ ANDROID_SDK_ROOT
```

各变量的路径修改为自己的工具包所在的路径，可以使用 export 命令来查看刚刚配置好的变量。如果没有列出刚刚配置好的变量，说明对 .bash_profile 文件的修改没有生效，因此在修改后，需要再执行一次 source .bash_profile 命令让修改立即生效。

3. 创建 CrossApp 工程

打开终端，输入如下命令创建工程。

```
cd $ CROSSAPP_ROOT
cd tools
cd project-creator
./create_project.py
```

输入以上命令后，会列出一个创建工程的示例，可以按照该示例，创建一个工程，例如：

```
./create_project.py - project MyFirstApp - package com.crossapp.test - language cpp
```

当提示 Have fun，表示工程创建成功，在 CrossApp 的 projects 目录下可以找到刚刚创建的工程。注意，不要直接复制以上命令到终端，有可能会出现格式错误，最好手动输出。

4. 导入 eclipse

（1）打开 eclipse，File→Other→Android→Android Project from Existing Code，单击 Browse，选择项目中的 porj.android，将 Android 工程导入 eclipse。

（2）工程导入 eclipse 后，首先需要将 CrossApp/CrossApp/platform/android/java/src 路径下的文件夹复制到工程的 /proj.android/src 下。

（3）右击工程→Properties→C/C++ Build→Environment，右侧单击 Add 添加一个变量，变量名为 NDK_ROOT，变量的值为 NDK 的根目录。

（4）以上两步完成后，连接 Android 手机到开发电脑，右击工程→Run As Android Application，开始编译工程。

> **注意** 以上步骤针对的是在 Mac 下创建的工程，如果是 Windows 下编译过的 CrossApp 工程转到 Mac 下调试，需要将 C/C++ Build 里面的 Build command 改为 bash ${ProjDirPath}/build_native.sh，同时还应该去除掉工程 jni/Android.mk 里面之前在 Windows 下额外添加的 import-add-path 相关路径。

在 Mac 下和 Windows 下配置 Android 开发环境的具体区别，请参阅不同环境下导入 eclipse 后的步骤。至此，Android 基本环境搭建完毕。

1.2.8 第一个 CrossApp 项目解析

打开 HelloCrossApp 下的 Classes 目录，可以看到如图 1-13 所示的结构。

图 1-13 在 Visual Studio 2013 中的 CrossApp 工程

Classes 目录包含了三个类。

（1）AppDelegate：项目入口类。

（2）FirstViewController：项目的视图控制器 CAViewController。

（3）RootWindow：项目启动的窗口 CAWindow。

1. CAWindow 介绍

CAWindow（窗口）主要的作用是作为所有 View（视图）的载体和容器，分发触摸消息，协同 ViewController 完成对应用程序的管理。应用程序通常只有一个 Window，即使存在多个 Window，也只能有一个 Window 能够接收屏幕事件。应用程序启动时创建这个窗口，并在窗口中加入一个或多个视图并显示出来，之后很少需要再次引用它。CAWindow 是所有 CAView（视图）的根，管理和协调应用程序的显示。

2. CAViewController 介绍

CAViewController（视图控制器）作为 CAView 的管理器，其最基本的功能就是控制视图的切换。视图控制器在 MVC 设计模式中扮演控制层（C）的角色，CAViewController 的作用就是管理与之关联的 View，同时与其他 CAViewController 相互通信和协调。

3. AppDelegate 是 CrossApp 的入口类

1）AppDelegate.h

打开 AppDelegate.h 文件，代码如下：

```
#ifndef _APP_DELEGATE_H_
#define _APP_DELEGATE_H_
#include "CrossApp.h"
class AppDelegate : private CrossApp::CCApplication
{
public:
    AppDelegate();
    virtual ~AppDelegate();
    /*
    这个函数用于实现 CAApplication 和 CAWindow(CCScene 应该为注释的错误，以后版本会修正)的初始化
    如果返回 true，则初始化成功，程序正常运行
    如果返回 false，则初始化失败，程序终止运行
    */
    virtual bool applicationDidFinishLaunching();
    //当程序进入后台运行时，此函数会被调用(例如，电话)
    virtual void applicationDidEnterBackground();
    //当程序从后台切回被激活时调用此函数
    virtual void applicationWillEnterForeground();
};
#endif // _APP_DELEGATE_H_
```

代码解析如下：

AppDelegate 继承 CCApplication 作为应用程序入口类，该类会在系统平台的 main 函数中创建实例，因此开发者只需要在该类中完成一款应用的生命周期管理即可，该类中 3 个重要的方法为

```
virtual bool applicationDidFinishLaunching();
virtual void applicationDidEnterBackground();
virtual void applicationWillEnterForeground();
```

开发者需要在这 3 个方法中编写正确的逻辑。

2）Application.cpp

打开 Application.cpp 代码如下：

```
bool AppDelegate::applicationDidFinishLaunching()
{
    //初始化应用程序
    CAApplication* pDirector = CAApplication::getApplication();
    //初始化绘图窗口
```

```cpp
    CCEGLView * pEGLView = CCEGLView::sharedOpenGLView();
    pDirector->setOpenGLView(pEGLView);
    //启动 RootWindow 入口
    pDirector->runWindow(RootWindow::create());
    return true;
}
void AppDelegate::applicationDidEnterBackground()
{
    //暂停动画
    CAApplication::getApplication()->stopAnimation();
    //暂停音效
    // SimpleAudioEngine::sharedEngine()->pauseBackgroundMusic();
}
void AppDelegate::applicationWillEnterForeground()
{   //恢复动画
    CAApplication::getApplication()->startAnimation();
    //恢复音效
    //SimpleAudioEngine::sharedEngine()->resumeBackgroundMusic();
}
```

代码解析如下：

AppDelegate.cpp 中 applicationDidFinishLaunching 方法定义了当应用启动完成以后初始化一个窗口，并显示这个窗口。

3）RootWindow.h

打开 RootWindow.h 文件，代码如下：

```cpp
class RootWindow: public CAWindow
{
public:
    static RootWindow * create();
    RootWindow();
    virtual ~RootWindow();
    virtual bool init();
};
#endif
```

RootWindow.Cpp 代码如下：

```cpp
#include "RootWindow.h"
#include "FirstViewController.h"
RootWindow * RootWindow::create()
{
    RootWindow * _window = new RootWindow();
    _window->init();
    _window->autorelease();
    return _window;
}
```

```cpp
bool RootWindow::init()
{
    if (!CAWindow::init())
    {   return false;
    }
    FirstViewController* _viewController = new FirstViewController();
    _viewController->init();
    this->setRootViewController(_viewController);
    _viewController->release();
    return true;
}
```

代码解析如下:

在该窗口中加载了一个视图控制器 FirstViewController,因此在屏幕上能看到的内容是在视图控制器中实现的初始化。

下面介绍 FirstViewController 中的初始化工作。

4) FirstViewController.h

打开 FirstViewController.h 文件,代码如下:

```cpp
#ifndef __HelloCpp__ViewController__
#define __HelloCpp__ViewController__
#include <iostream>
#include "CrossApp.h"
USING_NS_CC;
class FirstViewController: public CAViewController
{
public:
    FirstViewController();
    virtual ~FirstViewController();
protected:
    void viewDidLoad();
    void viewDidUnload();
};
#endif
```

代码解析如下:

在该视图控制器类中,主要包含了两个生命周期函数,分别定义该视图在加载和卸载时的逻辑。

```cpp
//加载函数一般做初始化 UI 和逻辑
void viewDidLoad();
//卸载函数一般做移除和释放
void viewDidUnload();
```

5) FirstViewController.cpp

打开 FirstViewController.cpp 文件,代码如下:

```cpp
void FirstViewController::viewDidLoad()
{
    CCRect winRect =
        this->getView()->getBounds(); //获得屏幕的Bounds(Bounds 不收缩放影响)
    //加载一张图片
    CAImageView * imageView =
        CAImageView::createWithImage(CAImage::create("HelloWorld.png"));
    //设置图片的 Frame 显示大小(Frame 的值缩放后会被改变)
    imageView->setFrame(winRect);
    //将图片添加的到屏幕上面
    //如果不添加,那么这个 UI 将不会被渲染,内存也会在下一个 loop 时被释放
    this->getView()->addSubview(imageView);
    //设置一个文本
    CALabel * label =
        CALabel::createWithCenter(CCRect(winRect.size.width * 0.5,
            winRect.size.height * 0.5 - 270, winRect.size.width, 200));
    //文本水平方向中间对齐
    label->setTextAlignment(CATextAlignmentCenter);
     //文本竖直方向中间对齐
    label->setVerticalTextAlignmet(CAVerticalTextAlignmentCenter);
    //设置字体大小(CROSSAPP_ADPTATION_RATIO 是用于适配的系数)
    label->setFontSize(72 * CROSSAPP_ADPTATION_RATIO);
    //设置文本内容
    label->setText("Hello World!");
    //设置文本颜色
    label->setColor(CAColor_white);
    //添加到屏幕并设置 z 轴
    this->getView()->insertSubview(label, 1);
}
```

代码解析如下:

通过 viewDidLoad 函数定义了在该视图控制器中包含的视图,一个图片视图,一个文字视图,这样就出现了在一个窗口(CAWindow)中包含了一个视图控制器(CAViewController),在该视图控制器中包含了两个 View。

下一章将介绍 CrossApp 中的核心概念,对于开发者而言,了解了第一个 CrossApp 的程序接口,可以快速上手来开发所需要的 App,因为在 CrossApp 中提供了很多不同类型的 View。

第 2 章 CrossApp 基础概念

2.1 核心类

CrossApp 是基于 MVC 模式的，MVC 指的是 Model（模型）、View（视图）和 Controller（控制），三者各司其职，共同构建一个完整的应用程序。

CrossApp 的核心类也是基于这样的设计思想来设计实现的。

通过本节学习 CAView、CAViewController 和 CAWindow 等能更好地了解 CrossApp 的 MVC 结构的原理与实现，能帮助我们更深刻地理解引擎的使用方法。

2.1.1 CAView

CAView 视图类是 CrossApp 引擎最基本的类，负责将各式各样的界面呈现出来，在 App 中所能看到的一切界面其实是一个或多个 CAView 组合构建的。CAView 负责在屏幕上定义矩形区域，在展示界面及响应用户界面交互方面发挥关键作用。每个 CAView 对象要负责渲染视图区域中的内容，并响应该区域内发生的操作事件，视图是应用程序与用户交互的重要机制。

除了显示内容和处理事件之外，CAView 还可以管理一个或多个子视图。可以在一个 CAView 上添加多个 CAView 或 CAView 的派生类，添加 CAView 的称为父节点，被添加的 CAView 被称为子节点。父节点负责管理其子节点，并根据需要调整它们的位置和尺寸，以及响应它们没有处理的事件。

CAView 常用的一些 public 函数如表 2-1 所示。

表 2-1 CAView 常用的 Public 函数

函　数	说　明
static CAView * create(void)	创建一个自动释放的 CAView
static CAView * createWithFrame(const CCRect& rect)	根据 Frame 创建 CAView
static CAView * createWithFrame(const CCRect& rect, const CAColor4B& color4B)	根据 Frame 和 Color 创建 CAView

续表

函 数	说 明
static CAView * createWithCenter(const CCRect& rect)	根据 Center 创建 CAView
static CAView * createWithCenter(const CCRect& rect, const CAColor4B& color4B)	根据 Center 和 Color 创建 CAView
static CAView * createWithColor(const CAColor4B& color4B)	根据 Color 创建 CAView
virtual bool init()	初始化
virtual bool initWithFrame(const CCRect& rect)	根据 Frame 初始化
virtual bool initWithFrame(const CCRect& rect, const CAColor4B& color4B)	根据 Frame 和 Color 初始化
virtual bool initWithCenter(const CCRect& rect)	根据 Center 初始化
virtual bool initWithCenter(const CCRect& rect, const CAColor4B& color4B)	根据 Center 和 Color 初始化
virtual bool initWithColor(const CAColor4B& color4B)	根据 Color 初始化
const char * description(void)	
virtual void setZOrder(int zOrder)	设置 Z 轴
virtual int getZOrder()	获得 Z 轴
virtual void setVertexZ(float vertexZ)	
virtual float getVertexZ()	
virtual void setScaleX(float fScaleX)	设置 X 方向的缩放因子
virtual float getScaleX()	获得 X 方向的缩放因子
virtual void setScaleY(float fScaleY)	设置 Y 方向的缩放因子
virtual float getScaleY()	获得 Y 方向的缩放因子
virtual void setScale(float scale)	设置缩放因子(包括 X 和 Y)
virtual float getScale();	获得缩放因子(包括 X 和 Y)
virtual void setScale(float fScaleX, float fScaleY)	设置缩放因子(第一个参数 X,第二个参数 Y)
virtual void setSkewX(float fSkewX)	设置 X 轴的扭曲因子
virtual float getSkewX()	获得 X 轴的扭曲因子
virtual void setSkewY(float fSkewY)	设置 Y 轴的扭曲因子
virtual float getSkewY()	获得 Y 轴的扭曲因子
virtual void setFrame(const CCRect& rect)	设置 Frame
virtual const CCRect& getFrame() const	获得 Frame
virtual void setFrameOrigin(const CCPoint& point)	设置 Frame 的起点
virtual const CCPoint& getFrameOrigin()	获得 Frame 的起点
virtual void setBounds(const CCRect& rect)	设置 Bounds
virtual CCRect getBounds() const	获得 Bounds
virtual void setCenter(CCRect rect);	设置 Center
virtual CCRect getCenter()	获得 Center
virtual void setCenterOrigin(const CCPoint& point)	设置 Center 的中心点

续表

函 数	说 明
virtual CCPoint getCenterOrigin()	获得 Center 的中心点
virtual void setVisible(bool visible)	设置显示或者隐藏
virtual bool isVisible()	是否显示状态
virtual void setRotation(float fRotation)	设置旋转角度
virtual float getRotation()	获得选择角度
virtual void setRotationX(float fRotaionX)	设置 X 轴旋转角度
virtual float getRotationX();	获得 X 轴旋转角度
virtual void setRotationY(float fRotationY)	设置 Y 轴旋转角度
virtual float getRotationY()	获得 Y 轴旋转角度
virtual void addSubview(CAView * child)	添加子节点
virtual void insertSubview(CAView * subview, int z)	添加子节点并设置 Z 轴
virtual CAView * getSubviewByTag(int tag)	根据 tag 值查找子节点
virtual CAView * getSubviewByTextTag(const std::string& textTag)	根据 testTag 值查找子节点
virtual const CAVector<CAView * >& getSubviews()	获得所有的子节点,范围为 CAVector 数组
virtual unsigned int getSubviewsCount(void) const	获得子节点的个数
virtual void setSuperview(CAView * superview)	设置父节点
virtual CAView * getSuperview()	获得父节点
virtual void removeFromSuperview()	把自己从父节点移除
virtual void removeSubview(CAView * subview)	移除子节点
virtual void removeSubviewByTag(int tag)	根据 tag 值移除子节点
virtual void removeSubviewByTextTag(const std::string& textTag)	根据 textTag 值移除子节点
virtual void removeAllSubviews()	移除所有子节点
virtual void reorderSubview(CAView * child, int zOrder)	重新在 Z 轴调整子节点
virtual void sortAllSubviews()	排序所有的子节点
virtual bool isRunning()	是否在运行
virtual void draw(void)	绘制
virtual CAView * copy()	复制

除了上述函数,CAView 还提供坐标系转换,关于坐标系的问题,将在后面的章节详细讲解。

2.1.2 CAViewController

CAViewController 作为 CAView 的管理类,其最基本的功能就是控制视图的切换。视图控制器在 MVC 设计模式中扮演控制层(Controller)的角色,CAViewController 的作用就是管理与之关联的 CAView,同时与其他 CAViewController 互相通信和协调。

CAViewController 的常用函数如表 2-2 和表 2-3 所示。

表 2-2 CAViewController 常用的 Public 函数

函　数	说　明
bool isViewRunning()	是否在运行
void setNavigationBarItem(CANavigationBarItem * item)	
void setTabBarItem(CATabBarItem * item)	设置 CATabBarItem
void presentModalViewController(CAViewController * controller, bool animated)	切换 CAViewController
void dismissModalViewController(bool animated)	释放
virtual bool isKeypadEnabled()	键盘是否启动
virtual void setKeypadEnabled(bool value)	设置键盘是否启动
virtual CAView * getView()	获得当前的 View

表 2-3 CAViewController 常用的 Protected 函数

函　数	说　明
virtual void viewDidLoad()	加载时调用
virtual void viewDidUnload()	释放时候调用

2.1.3　CAWindow

CAWindow 是所有 CAView 的载体，并且负责分发触摸消息，协同 CAViewController 完成对应程序的管理。应用程序通常只有一个 CAWindow，即使存在多个 CAWindow 也只有一个 CAWindow 能接收屏幕事件。应用程序启动时创建一个 CAWindow，并将多个 CAView 添加到 CAWindow 并显示出来，之后就很少再引用它。CAWindow 是屏幕上显示的 CAView 的根节点。

2.2　内存管理

CrossApp 的内存管理移植自 Objective-C，没有接触过 OC 的 C++开发人员可能对此有些迷惑。不过只要明白其中原理，也是很容易理解和使用的。

CrossApp 通过 CAObject 和 CAPoolManager 来实现内存管理。所有使用 CrossApp 引用记数机制的类都必须派生自 CAObject。CAObject 有一个记数器成员变量 m_uReference，当 CAObject 被构造时 m_uReference=1，表示该对象被引用 1 次。

CAObject 的 retain 方法可以使记数器加 1，release 方法可以使记数器减 1。当记数器减至 0 时 release 方法会通过 delete this 来销毁自己。

2.2.1　对象内存引用记数

CrossApp 内存管理的基本原理就是对象内存引用记数，CrossApp 将内存引用记数的实现放在顶层父类 CAObject 中，这里将涉及引用记数的 CAObject 的成员和方法摘录如下：

```cpp
class CC_DLL CAObject : public CACopying
{
public:
    unsigned int        m_uID;
protected:
    unsigned int        m_uReference;
    unsigned int        m_uAutoReleaseCount;
public:
    void release(void);
    CAObject* retain(void);
    CAObject* autorelease(void);
};
CAObject::CAObject(void)
: m_uReference(1)
, m_uAutoReleaseCount(0)
, m_nTag(kCAObjectTagInvalid)
{
    static unsigned int uObjectCount = 0;
    m_uID = ++uObjectCount;
}
CAObject::~CAObject(void)
{
    CAScheduler::unscheduleAllForTarget(this);
    if (m_uAutoReleaseCount > 0)
    {
        CAPoolManager::sharedPoolManager()->removeObject(this);
    }
    CCScriptEngineProtocol* pEngine =
        CCScriptEngineManager::sharedManager()->getScriptEngine();
    if (pEngine != NULL && pEngine->getScriptType() == kScriptTypeJavascript)
    {
        pEngine->removeScriptObjectByCCObject(this);
    }
    void CAObject::release(void)
    {
        --m_uReference;
        if (m_uReference == 0)
        {
            delete this;
        }
    }
    CAObject* CAObject::retain(void)
    {
        ++m_uReference;
```

```
        return this;
    }
    CAObject * CAObject::autorelease(void)
    {
        CAPoolManager::sharedPoolManager()->addObject(this);
        return this;
    }
```

代码解析如下：

通过以上代码可以看到，对象记数的核心字段是 m_uReference。
(1) 当一个 Object 初始化(被 new 出来时)，m_uReference = 1；
(2) 当调用该 Object 的 retain 方法时，m_uReference++；
(3) 当调用该 Object 的 release 方法时，m_uReference--，
(4) 若 m_uReference 减至 0，则 delete 该 Object。

2.2.2　手工对象内存管理

在上述对象内存引用记数的原理下，得出在 Cocos2d-x 下手工对象内存管理的基本模式：

```
CCObject * obj = new CCObject();
//初始化
obj->init();
//释放
obj->release();
```

2.2.3　自动对象内存管理

自动对象内存管理指的是那些不再需要的 object 将由 CrossApp 引擎释放，而无须手工调用 Release 方法。

自动对象内存管理显然也要遵循内存引用记数规则，只有当 object 的记数变为 0 时，才会释放对象的内存。

自动对象内存管理的典型模式如下：

```
CAMyClass * CAMyClass::create()
{
    CAMyClass * pRet = new CAMyClass();
    if (pRet && pRet->init())
    {
        pRet->autorelease();
        return pRet;
    }
    else
    {
```

```
        CC_SAFE_DELETE(pRet);
        return NULL;
    }
}
```

一般通过一个单例模式创建对象,与手工模式的不同之处在于 init 后多了一个 autorelease 调用。这里再把 autorelease 调用的实现摘录如下:

```
CAObject * CAObject::autorelease(void)
{
    CAPoolManager::sharedPoolManager()->addObject(this);
    return this;
}
```
追溯 addObject 方法:
```
//CAPoolManager.cpp 中的方法
void CAPoolManager::addObject(CAObject * pObject)
{
    getCurReleasePool()->addObject(pObject);
}
// CAAutoreleasePool.cpp 中的方法
void CAAutoreleasePool::addObject(CAObject * pObject)
{
    m_pManagedObjectArray->addObject(pObject);
    CCAssert(pObject->m_uReference > 1, "reference count should be greater than 1");
    ++(pObject->m_uAutoReleaseCount);
    pObject->release(); // no ref count, in this case autorelease pool added.
}
//CCArray.cpp 中的方法
void CCArray::addObject(CAObject * object)
{
    ccArrayAppendObjectWithResize(data, object);
}
// ccCArray.cpp 中的方法
void ccArrayAppendObjectWithResize(ccArray * arr, CAObject * object)
{
    ccArrayEnsureExtraCapacity(arr, 1);
    ccArrayAppendObject(arr, object);
}
void ccArrayAppendObject(ccArray * arr, CAObject * object)
{
    CCAssert(object != NULL, "Invalid parameter!");
    object->retain();
    arr->arr[arr->num] = object;
    arr->num++;
}
```

调用层次深,涉及的类也多,这里进行归纳总结。

CrossApp 的自动对象内存管理基于对象引用记数以及 CAAutoreleasePool（自动释放池）。引用记数前面已经介绍，此处不再赘述，只对自动释放池进行说明。CrossApp 关于自动对象内存管理的基本类层次结构如下：

（1）CAPoolManager 类（自动释放池管理器）

```
-CCArray*      m_pReleasePoolStack;
//（自动释放池栈,存放 CCAutoreleasePool 类实例)
```

（2）CAAutoreleasePool 类

```
-CCArray*      m_pManagedObjectArray;
//（受管对象数组）
```

CAObject 关于内存记数以及自动管理有两个字段：m_uReference 和 m_uAutoReleaseCount。前面在手工管理模式下，只提及了 m_uReference，现在介绍 m_uAutoReleaseCount。可以根据自动释放对象的创建步骤来理解不同阶段下这两个重要字段的值及其含义。

```
CCYourClass* pRet = new CCYourClass();    // m_uReference = 1; m_uAutoReleaseCount = 0;
pRet->init();                             //  m_uReference = 1; m_uAutoReleaseCount = 0;
pRet->autorelease();
```

详细分解一下 pRet->autorelease() 在 CAAutoreleasePool::addObject 中执行步骤如下：

```
m_pManagedObjectArray->addObject(pObject);// m_uReference = 2 m_uAutoReleaseCount = 0
++(pObject->m_uAutoReleaseCount);         //m_uReference = 2  m_uAutoReleaseCount = 1
pObject->release();                       //m_uReference = 1  m_uAutoReleaseCount = 1
```

在调用 autorelease 之前，这两个值与手工模式并无差别，在 autorelease 后，m_uReference 值没有变，但 m_uAutoReleaseCount 被加 1。

m_uAutoReleaseCount 这个字段的名字很容易让人误以为是个记数器，但实际上绝大多数时刻它是一个标识的角色，以前版本代码中有一个布尔字段 m_bManaged，但后来被 m_uAutoReleaseCount 替换了，因此 m_uAutoReleaseCount 兼有 m_bManaged 的含义，即该 object 是否在自动释放池的控制之下。如果在自动释放池的控制下，自动释放池会定期调用该 object 的 release 方法，直到该 object 内存记数降为 0，被真正释放；否则该 object 不能被自动释放池自动释放内存，需手工释放。理解该点非常重要，在后面在项目开发过程中如果不希望异常退出必然要理解该原理。

2.3 坐标系

加深理解基础概念能够大大提高学习 CrossApp 的效率。本小节将简单介绍 CrossApp 的坐标系统。

基础坐标系 CrossApp 采用的坐标系是屏幕坐标系，即左上角为原点，向右对应 X 轴增长方向，向下对应 Y 轴增长方向，如图 2-1 所示。

图 2-1　CrossApp 坐标系

下面我们介绍节点的概念。

谈到 CrossApp 的坐标系，不得不谈到视图类 CAView。视图类 CAView 是整个 CrossApp 引擎最基本的类，负责将各式各样的界面呈现出来，我们在 App 中所能看见的一切界面其实就是一个个 CAView 的组合。

CAView 负责在屏幕上定义矩形区域，在展示用户界面及响应用户界面交互方面发挥关键作用。每个视图对象要负责渲染试图矩形区域中的内容，并响应该区域内发生的操作事件，视图是应用程序用户交互的重要机制。

除了显示内容和处理事件之外，视图还可以管理一个或多个子视图。

我们可以在一个视图上面添加多个子视图，而作为父视图，即父节点，负责管理其直接子视图，并根据需要调整它们的位置和尺寸，以及响应它们没有处理的事件。

根据类说明可以得出以下结论：

（1）所有能看到的都是 CAView 的派生类；

（2）CAView 上面可以添加子 CAView；

（3）父节点管理子视图。

初学者可能看不明白父节点和子视图的概念，并且 B 是添加在 A 上的。假如这时候调整 A 的坐标位置，那么 B 也将随 A 的坐标改变而改变。

这样我们就说：B 是 A 的子节点（也称子视图），A 是 B 的父节点。

由于 B 是 A 的子节点（子视图），那么 B 可以使用 A 的节点坐标系，如图 2-2 所示。

如图 2-3 所示，屏幕上显示了两个 CAView 分别是 A 和 B 的起始点、中心点、宽和高。

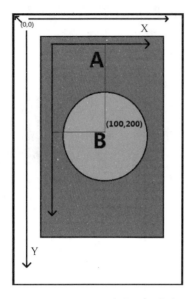
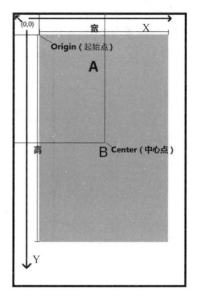

图 2-2　CrossApp 中的父子节点　　　图 2-3　CrossApp 中节点属性

CAView 在坐标系中分别定义了三个属性：Frame、Bounds 和 Center。分别介绍如下。

1. Frame

类型：CCRect。

确定 View 在屏幕上显示的位置和大小，参考的是父视图的坐标系统，Frame 属性是 View 及其子类共有属性。CCRect 包括两个成员，一个是起点坐标 Origin；另一个是宽高 size。创建一个 View 必须指定 Frame，否则没有任何效果。

对于 CAView 及其子类的 Frame 操作，在进行 createWithFrame 和 setFrame 等操作时，如果 CCRect 的 size 成员设置为(0,0)，则不改变 View 的大小；如果只想改变 View 的坐标，不需要改变 View 的大小，则可采用 setFrameOrigin 方法来进行设置。

2. Bounds

类型：CCRect。

View 在自身坐标系统中的位置和大小，参考的是自身的坐标系统，CCRect 的 Origin 值永远都是(0,0)，Bounds 属性是 View 及其子类的共有属性。

在设置 Frame 时，Bounds 也确定，其值等于 Frame 的值。Bounds 和 Frame 是有区别的，Frame 的值会随着 View 的缩放操作改变，Bounds 的值则不会改变。

3. Center

类型：CCRect。

表示 View 的中心点在屏幕上的位置。CrossApp 采用的坐标系是屏幕坐标系，即左上角为原点，向右和向下分别对应 X 和 Y 轴。在 CrossApp 中确定一个 View 的位置，是通过

Origin 和 size 来设定的。为了方便,这里可以直接使用 Center 将 View 的中心点设置在想要的位置。如果只想改变 View 的坐标,不需要改变 View 的大小,则可采用 setFrameOrigin 方法来进行设置。

以上这些属性都是针对当前节点的父节点进行定义的。

2.4 适配方案

屏幕适配没有绝对完美的方案,不可能做到一套适配方案适用所有设备。

当一个 App 运行在不同设备时,由于设备的物理尺寸和分辨率不尽相同,所以某个界面里的一些 UI 元素可能会发生变化,很多情况下需要开发者自己去控制,并且在程序中我们应尽量使用相对坐标。

我们的 App 运行到 PPI 较低的设备时,界面可能会被放大,当运行到 PPI 较高的设备时界面又可能会被缩小。但实际上,我们希望某些 UI 元素保持不变,因此对于某些大小锁定的 UI 元素来说,需要固定它在不同设备上显示都保持一样的大小,比如小说阅读器里的文本字体大小,并且对于屏幕尺寸大的设备应该显示更多的内容,例如可以固定 tableView 的每个 cell 的大小,从而在尺寸较大的屏幕上显示更多的 cell。对此,CrossApp 提供了一种辅助的适配方案。

以 iPhone4 和 iPhone5 为基准进行适配,表 2-4 展示出了相关参数。

表 2-4　CrossApp 中的适配参数

设　　备	屏幕尺寸	分　辨　率	PPI
iPhone4	3.4 英寸	960 * 640	326
iPhone5	4.0 英寸	1136 * 640	326

CrossApp 辅助适配方案适配的原则是尽量保持 UI 的物理尺寸相近,但仍旧可能会有细微变化,具体的变化值由系统决定。在不同设备下,PPI 值会有所不同,在保持 UI 的物理尺寸不变的情况下,在不同设备上的显示效果会有所差别。屏幕尺寸大的设备,显示的内容会相对较多。PPI 值越高,则画面越精细。

这里有一个修正系数的概念,因为以 iPhone4 和 iPhone5 为基准,所以 iPhone4 和 iPhone5 的修正系数为 1,但对于其他设备,该系数是个近似值,当我们设置 UI 的 size 时,如果需要保持其在不同设备上的显示大小一样,通常需要乘以该修正系数,CrossApp 会根据不同设备的规格自动计算该修正系数的值。

使用时只需要把设定的数值传递给内联函数 _px 就可以保证在不同设备上的 UI 物理尺寸不变。例如,设定 button 的物理尺寸为 100 * 50。

```
CADipSize size = typeView->getBounds().size;
CAButton* squareRectBtn = CAButton::create(CAButtonTypeSquareRect);
squareRectBtn->setCenter(CADipRect(300, 200, 100, 50));
```

```
squareRectBtn->setTitleForState(CAControlStateAll, "SquareRect");
this->getView()->addSubview(squareRectBtn);
```

这里简单介绍 PPI 和 DPI 的概念。

（1）PPI：每英寸所包含的像素。在 iOS 设备上，通常以 PPI 来表示设备屏幕的显示精细度。

（2）DPI：每英寸所包含的点。在 Android 设备上，通常以 DPI 来表示设备屏幕的显示精细度。

通常情况下，PPI 和 DPI 可以表示同一个概念，PPI 主要针对显示设备，DPI 更多应用于打印输出设备，但都表示每英寸所包含的像素点。

2.5 深入理解 CAViewController 和 MVC

CrossApp 是基于 MVC 模式的，这里的 MVC 即模型层（Model）、视图层（View）、控制层（Controller），三者各司其职，共同维持一个完整的应用程序。

关于 MVC 模式，并没有一个非常明确的概念，在不同的开发环境中设计可能有所区别。面向对象的目标就是设计出低耦合、高聚合的程序，MVC 模式提供了一种很好的解决方案。

在 CrossApp 中的 Model 层也就是我们各种数据原型、业务逻辑和算法，View 层顾名思义就是展现给用户的 UI 界面，而 Controller 层的职责就是把 Model 和 View 两个相互关联不大的层衔接起来。CrossApp 通过 CAView 和丰富的 UI 控件来实现 View 层的功能，通过 CAViewController 及其子类来共同完成 Controller 层，实现 Model 和 View 之间的通信。

View 层和 Model 层通常不能直接进行通信，View 层接收用户的操作，并把操作通知给 Controller，Controller 接收到消息后就更新 Model 层。同理，Model 层的数据发生变动后，通知 Controller，Controller 提示 View 层进行界面更新。

2.5.1 CAViewController 的职责

CAViewController 负责监听键盘，例如 Android 平台的 Back 键和 Menu 键或 PC 端的键盘输入。CAViewController 中的按键回调如表 2-5 所示。

表 2-5　CAViewController 中的按键回调

函　　数	说　　明
virtual bool isKeypadEnabled()	是否监听物理键（如 android 的 Back 键或 pc 的 Esc 键等）
virtual void setKeypadEnabled(bool value)	设置是否监听物理返回键，默认为 false
virtual void keyBackClicked()	back 键响应函数（android 平台）
virtual void keyMenuClicked()	menu 键响应函数（android 平台）

CAViewController 还负责接送屏幕触摸的交互：触摸开始、触摸移动、触摸抬起、触摸中断（非正常结束，如突然来电话了）等状态。CAViewController 中的触屏回调如表 2-6 所示。

表 2-6 CAViewController 中的触屏回调

函 数	说 明
virtual bool ccTouchBegan(CATouch * pTouch, CAEvent * pEvent)	触摸开始
virtual void ccTouchMoved(CATouch * pTouch, CAEvent * pEvent)	触摸移动
virtual void ccTouchEnded(CATouch * pTouch, CAEvent * pEvent)	触摸结束
virtual void ccTouchCancelled(CATouch * pTouch, CAEvent * pEvent)	触摸非正常结束（如电话或锁屏）

CAViewController 还负责视图切换管理，CrossApp 系统为我提供了四种视图管理器（见图 2-4、图 2-5）：

(1) CAViewController 基础视图管理器：没有额外样式。
(2) CATabBarController 标签视图管理器：顶部（底部）多排标签栏，用于切换。
(3) CANavigationController 导航视图管理器：顶部（底部）多一个导航栏。
(4) CADrawerController 抽屉视图管理器：横向滑动，显示隐藏界面。

图 2-4 CrossApp 中基础视图管理器和导航视图管理器

图 2-5 CrossApp 中标签视图管理器和抽屉视图管理器

2.6 CAViewController 类的使用

表 2-7 是 CAViewController 类的常用的函数。

表 2-7 CAViewController 中的常用函数

函　　数	说　　明
virtual bool init()	初始化,默认返回 true,如果返回 false 则初始化失败
bool isViewRunning()	当前 View 是否在运行
Void setNavigationBarItem(ANavigationBarItem * item)	设置 CANavigationBarItem 属性(只有被 CANavigation Controller 管理才会显示)
void setTabBarItem(CATabBarItem * item)	设置 CATabBarItem 属性(只有被 CATableController 管理才会显示)
void presentModalViewController(CAViewController * controller, bool animated);	弹出一个 CAViewController controller：要弹出的 animated：是否需要弹出动画
void dismissModalViewController(bool animated)	移除 CAViewController animated：是否需要动画

续表

函 数	说 明
virtual bool isKeypadEnabled()	是否监听物理键(android 的 Back 键或 pc 的 Esc 键)
virtual void setKeypadEnabled(bool value)	设置是否监听物理返回键,默认 false value：true 允许,false 禁止
virtual CAView* getView()	当前 CAViewController 的 View 根节点
virtual CAResponder* nextResponder()	获得下个监听者
virtual bool ccTouchBegan(CATouch* pTouch, CAEvent* pEvent)	触摸开始
virtual void ccTouchMoved(CATouch* pTouch, CAEvent* pEvent)	触摸移动
virtual void ccTouchEnded(CATouch* pTouch, CAEvent* pEvent)	触摸结束
virtual void ccTouchCancelled(CATouch* pTouch, CAEvent* pEvent)	触摸非正常结束(例如：电话或锁屏)
virtual void viewDidLoad()	当前控制器的 view 被加载完毕后调用
virtual void viewDidUnload()	当前控制器的 view 被移除掉时调用
virtual void viewDidAppear()	显示时被调用
virtual void viewDidDisappear()	隐藏时被调用
virtual void setKeypadEnabled(bool value)	设置是否监听物理返回键,默认为 false
virtual void keyBackClicked()	back 键响应函数(android 平台)
virtual void keyMenuClicked()	menu 键响应函数(android 平台)

2.6.1 CAViewController 生命周期

CAViewController 的生命周期是一个很重要的概念,它能帮助我们理解 CrossApp 的调用机制。理解这些调用,就能更好地管理内存及事件的触发。

viewDidLoad()——第一次被加载时调用(仅有一次会被调用)。

viewDidAppear()——显示时被调用(例如 TableController 切换到显示)。

viewDidDisappear()——隐藏时被调用(切换到其他 ViewController)。

viewDidUnload()——移除时被调用(仅有一次会被调用)。

注意 使用视图 CAWindow 对象的 RootWindow 直接加载的 CAViewController 对象显示时不会调用 viewDidAppear()函数。

2.6.2 CAViewController 使用

其实在前面的第一个 CrossApp 项目中,我们已经通过 FirstViewController 了解一些使用的方法。逻辑部分和 UI 的加载主要是在 viewDidLoad()中实现的。如果我希望监听

back 键,那么我们需要在 FirstViewController 重写 keyBackClicked() 函数并在函数中添加我们想要的逻辑,例如返回上一个界面,退出程序等。

监听触摸事件也是一样的道理。

如果我们想在当前的 ViewController 上面弹出另一个 ViewController,那么我就需要调用 presentModalViewController(CAViewController * controller, bool animated) 函数,并将要弹出的 ViewController 当成参数传入函数内。如果希望关闭,则需要调用新弹出窗口的 void dismissModalViewController(bool animated) 函数。

下面看一下代码演示。

首先新建一个 CopybookViewController 与 CAViewController,代码类似于 FirstViewController。
CopybookViewController.h 代码如下:

```cpp
#include <iostream>
#include "CrossApp.h"
USING_NS_CC;
class CopybookViewController: public CAViewController
{
public:
    CopybookViewController();
    virtual ~CopybookViewController();
    void Btncallback(CAControl *, CCPoint);
protected:
    void viewDidLoad();
    void viewDidUnload();
    void viewDidAppear();     //显示时被调用(例如 TableController 切换到显示)
    void viewDidDisappear();  //隐藏时被调用(例如 TableController 切换到其他 Controller
};
```

然后在 cpp 文件中添加一个 CAButton 按钮并绑定回调函数 Btncallback,在生命周期的几个函数中都添加上 log 打印,代码如下:

```cpp
CopybookViewController.cpp
#include "CopybookViewController.h"
CopybookViewController::CopybookViewController()
{
}
CopybookViewController::~CopybookViewController()
{
}
void CopybookViewController::viewDidLoad()
{
    CCLog("CopybookViewController::viewDidLoad--->");
    //Do any additional setup after loading the view from its nib.
    CCRect winRect = this->getView()->getBounds();
    CAImageView* imageView =
```

```cpp
        CAImageView::createWithImage(CAImage::create("r/HelloWorld.png"));
    imageView->setImageViewScaleType(CAImageViewScaleTypeFitImageCrop);
    imageView->setFrame(winRect);
    this->getView()->addSubview(imageView);
    CCLog("%f", CAApplication::getApplication()->getWinSize().width);
    //移除按钮
    CAButton* btn =
        CAButton::createWithCenter(CADipRect(400, 400, 400, 100),
        CAButtonTypeSquareRect);
    btn->setTitleForState(CAControlStateAll, "dismissModalViewController");
    btn->addTarget(this, CAControl_selector(CopybookViewController::Btncallback),
        CAControlEventTouchDown);
    this->getView()->addSubview(btn);
}
void CopybookViewController::Btncallback(CAControl*, CCPoint)
{
    //移除当前
    this->dismissModalViewController(true);
}
void CopybookViewController::viewDidUnload()
{   //Release any retained subviews of the main view.
    //e.g. self.myOutlet = nil;
    //移除时被调用(仅有一次会被调用)
    CCLog("CopybookViewController::viewDidUnload --->");
}
void CopybookViewController::viewDidAppear() //显示时被调用{
    CCLog("CopybookViewController::viewDidAppear --->");
}
void CopybookViewController::viewDidDisappear() //隐藏时被调用{
    CCLog("CopybookViewController::viewDidDisappear --->");
}
```

同样我们也改装 FirstViewController 中的代码,首先在头文件中添加

```cpp
FirstViewController.h
void Btncallback(CAControl*, CCPoint);
void viewDidLoad();
void viewDidUnload();
void viewDidAppear();      //显示时被调用(例如 TableController 切换到显示)
void viewDidDisappear();   //隐藏时被调用(例如 TableController 切换到其他 Controller
```

然后在 FirstViewController.cpp 添加一个 CAButton 并绑定监听函数:

```cpp
void FirstViewController::viewDidLoad()
{
    CCLog("FirstViewController::viewDidLoad --->");
    //Do any additional setup after loading the view from its nib.
    CCRect winRect = this->getView()->getBounds();
```

```cpp
        CAImageView* imageView = 
        CAImageView::createWithImage(CAImage::create("r/HelloWorld.png"));
            imageView->setImageViewScaleType(CAImageViewScaleTypeFitImageCrop);
        imageView->setFrame(winRect);
        this->getView()->addSubview(imageView);
        CCLog("%f",CAApplication::getApplication()->getWinSize().width);
    //弹出
        CAButton* btn = CAButton::createWithCenter(CADipRect(400, 400, 400, 100),
           CAButtonTypeSquareRect);
    btn->setTitleForState(CAControlStateAll, "presentModalView");
    btn->addTarget(this, CAControl_selector(FirstViewController::Btncallback),
        CAControlEventTouchDown);
    this->getView()->addSubview(btn);
}
void FirstViewController::Btncallback(CAControl*, CCPoint)
{
    //初始化
    CopybookViewController* cp = new CopybookViewController();
    //调用 init
    cp->init();
    //弹出
    this->presentModalViewController(cp,true);
    //释放内存
    cp->release();
}
void FirstViewController::viewDidUnload()
{
    // Release any retained subviews of the main view.
    // e.g. self.myOutlet = nil;
    //移除时被调用(仅有一次会被调用)
    CCLog("FirstViewController::viewDidUnload--->");
}
void FirstViewController::viewDidAppear() //显示时被调用(例如 TableController 切换到显示)
{
    CCLog("FirstViewController::viewDidAppear--->");
}
void FirstViewController::viewDidDisappear() //隐藏时被调用(例如 TableController 切换到其他 Controller
{
    CCLog("FirstViewController::viewDidDisappear--->");
}
```

这样就可以实现了,弹出新的 ViewController 并关闭它,我们观察输出窗口验证了我们生命周期的结论(见图 2-6)。

```
libpng warning: iCCP: known incorrect sRGB profile
2015-09-17 17:02:24.707 MyViewController[4857:178271]
CrossApp: surface size: 750x1294
CrossApp: FirstViewController::viewDidLoad--->
libpng warning: iCCP: known incorrect sRGB profile
CrossApp: 750.000000
CrossApp: Get data from file(/System/Library/Fonts/Cache/
STHeiti-Light.ttc) failed!
CrossApp: Get data from file(/System/Library/Fonts/Core/
STHeiti-Light.ttc) failed!
CrossApp: CopybookViewController::viewDidLoad--->
CrossApp: 750.000000
CrossApp: CopybookViewController::viewDidAppear--->
CrossApp: FirstViewController::viewDidDisappear--->

All Output ≎
```

图 2-6　CrossApp 中控制器的生命周期

2.7　CANavigationController 导航视图控制器

CANavigationController 是 CAViewController 的子类，它的作用是管理多个 CAViewController，我们要明白的是 CANavigationController 是使用堆栈的方式管理，即我们每往 CANavigationController 添加一个 CAViewController，则进行一次堆栈操作，每次移除则进行一次出栈操作。

常用函数如表 2-8 所示。

表 2-8　CANavigationController 中的常用函数

函　数	说　明
virtual bool initWithRootViewController (CAViewController * viewController, CABarVerticalAlignment var = CABarVerticalAlignmentTop)	使用 CAViewController 来初始化，这个是必须的 viewController：用来初始化 CAViewController，它将被 CANavigationController 压栈 var：CANavigationBar 的现实样式 CABarVerticalAlignmentTop：顶部显示， CABarVerticalAlignmentBottom，底部显示
virtual void replaceViewController (CAViewController * viewController, bool animated)	替换栈顶的 viewController viewController：新的 viewController animated：是否播放动画
virtual void pushViewController (CAViewController * viewController, bool animated)	设置 CANavigationBarItem 属性(只有被 CANavigationController 管理才会显示)
CAViewController * popViewControllerAnimated(bool animated)	移除栈顶的 viewController animated：是否播放动画
CAViewController * getViewControllerAtIndex(int index)	根据索引值获取 viewController

续表

函　　数	说　　明
CAViewController * getBackViewController()	返回最后一个 ViewController
inline unsigned long getViewControllerCount() {return m_pViewControllers.size()	当前栈内 viewController 总数
virtual void setNavigationBarHidden (bool hidden, bool animated)	是否隐藏 navigationBar

尝试一下如何使用 CANavigationController 来进行管理 ViewController，我们只需要修改 RootWindow 代码如下：

```
bool RootWindow::init()
{
    if (!CAWindow::init())
    {
        return false;
    }
    //创建 Navigation
    CANavigationController* _viewController = new CANavigationController();
    //创建 Navigation 的第一个 Controller
    FirstViewController* first = new FirstViewController();
    first->init();
    //使用一个 controller 初始化 Navigation(必须)
    _viewController->initWithRootViewController(first);
    //RootWindow 加载 Navigation
    this->setRootViewController(_viewController);
    //释放内存
    first->release();
    //释放内存
    _viewController->release();
    return true;
}
```

看看 FirstViewController 是不是不一样了——多了一个导航栏。导航栏上的文本是默认文本，我们可以设置我们想要的文本内容。看下面的代码：

```
bool RootWindow::init()
{
    if (!CAWindow::init())
    {
        return false;
    }
    //创建 Navigation
    CANavigationController* _viewController = new CANavigationController();
    //创建 Navigation 的第一个 Controller
```

```cpp
    FirstViewController* first = new FirstViewController();
    first->init();
    //创建CANavigationBarItem并设置显示标题
    CANavigationBarItem* nItem = CANavigationBarItem::create("First");
    //创建左边按钮(右边按钮同理)
    CABarButtonItem* leftBtn =
    CABarButtonItem::create("", CAImage::create("source_material/btn_left_white.png"),
    CAImage::create("source_material/btn_left_blue.png"));
    //将leftBtn添加到CANavigationBarItem
    nItem->addLeftButtonItem(leftBtn);
    //将CANavigationBarItem添加到FirstViewController
    first->setNavigationBarItem(nItem);
    //使用一个controller初始化Navigation(必须)
    //CABarVerticalAlignmentBottom显示在底部
    _viewController->initWithRootViewController(first,CABarVerticalAlignmentBottom);
    //RootWindow加载Navigation
    this->setRootViewController(_viewController);
    //释放内存
    first->release();
    //释放内存
    _viewController->release();
    return true;
}
```

我们可以改变导航栏的文本、背景色、图片及添加左按钮和右按钮等样式，还可以通过向CANavigationController栈内添加或删除ViewController进行界面的管理和切换。比如我们为左按钮绑定一个回调函数：

```cpp
leftBtn->setTarget(this, CAControl_selector(RootWindow::callback));
```

在callback函数中实现：

```cpp
void RootWindow::callback(CAControl*, CCPoint)
{
    SecondViewController* second = new SecondViewController();
    second->init();
    _viewController->pushViewController(second, true);
    second->release();
}
```

这样当单击左按钮时，界面就会跳转到SecondViewController。

2.8 CATabBarController 切换视图控制器

CATabBarController作为一个特殊的视图管理器(常用函数见表2-9)，负责协调多个视图管理器之间的工作，是对视图管理器的一种特殊封装。通常当你的程序需要使用一些平行的界面(这里说的平行界面就是程序中的某些功能界面是处于平级的),这些功能界面可以相互切换,tabBarController 就很适合这种情况。

表 2-9　CANavigationController 中的常用函数

函　　数	说　　明
virtual bool initWithViewControllers(const CAVector<CAViewController *> & viewControllers,CABarVerticalAlignment var = CABarVerticalAlignmentBottom)	初始 CATabBar viewControllers:含有 ViewController 的数组 CABarVerticalAlignment:切换条位置(上部,下部)
bool showSelectedViewController (CAViewController * viewController)	设置当前被选中的 viewController viewController:要设置选中的指针
CAViewController * getViewControllerAtIndex(unsigned int index)	获取当前显示 viewController 的索引值 index:索引位置(从 0 开始)
CAViewController * getSelectedViewController()	获取当前选中的 viewController
virtual bool showSelectedViewControllerAtIndex(unsigned int index)	根据索引值显示当前选中的 viewController index:索引位置(从 0 开始)
virtual unsigned int getSelectedViewControllerAtIndex()	获取当前的被选中的 viewController 的索引值
virtual void setTabBarHidden(bool hidden, bool animated)	设置 TabBar 的隐藏与 hidden:是否隐藏(默认 false) animated:是否开启动画效果
void updateItem (CAViewController * viewController)	更新视图 viewController:需要更新的视图管理器
void showTabBarSelectedIndicator()	显示刷新 TabBar

我们看看代码部分是如何实现的:

```
bool RootWindow::init()
{
    if (!CAWindow::init())
    {
        return false;
    }

    FirstViewController * first = new FirstViewController();
    first->init();
    first->setTabBarItem(CATabBarItem::create(
        UTF8("第一项"),CAImage::create(""),CAImage::create("")));
    SecondViewController * Second = new SecondViewController();
    Second->init();
    Second->setTabBarItem(CATabBarItem::create(
```

```
        UTF8("第二项"), CAImage::create(""), CAImage::create("")));
ThirdViewController * Third = new ThirdViewController();
Third->init();
Third->setTabBarItem(CATabBarItem::create(
        UTF8("第三项"), CAImage::create(""), CAImage::create("")));
//将多个 ViewController 放到 CAVector 进行管理
CAVector<CAViewController *> vector;
vector.pushBack(first);
vector.pushBack(Second);
vector.pushBack(Third);
//创建 TabBar
CATabBarController * tab = new CATabBarController();
//通过含有 ViewControler 的 CAVector 进行初始化
tab->initWithViewControllers(vector);
//设置可以滑动切换
tab->setScrollEnabled(true);
tab->showTabBarSelectedIndicator();
this->setRootViewController(tab);
//是否内存
first->release();
Second->release();
Third->release();
tab->release();
return true;
}
```

CrossApp 中的 Tab 视图控制器如图 2-7 所示。

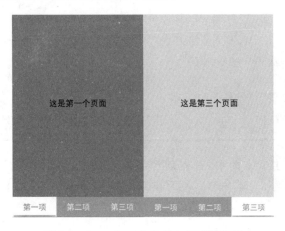

图 2-7　CrossApp 中的 Tab 视图控制器

同样，我们也可以修改它的样式：文本、位置(上、下)、背景颜色、图片。
CATabBarItem 控制：文本、默认图片、选中图片。

```
//title: 文本,image: 默认背景图片,selectedImage: 选中时背景图片
```

```
static CATabBarItem * create(const std::string& title, CAImage * image, CAImage * selectedImage = 
NULL);
```

CATabBarController 控制：显示位置（初始时设置）、背景颜色、图片。

```
//初始时设置显示位置在顶部
tab->initWithViewControllers(vector,CABarVerticalAlignmentTop);
//是否隐藏 TabBar
tab->setTabBarHidden(false, false);
//默认背景颜色
tab->setTabBarBackGroundColor(CAColor_orange);
//默认背景图片
tab->setTabBarBackGroundImage(CAImage::create("source_material/btn_left_blue.png"));
```

2.9 CADrawerController 侧边抽屉式导航控制器

CADrawerController 是易用的侧边抽屉式导航控制器（常用函数见表 2-6）。它能同时控制两个视图，横向滑动时，左边被隐藏的视图会像拉抽屉一样展示出来。

表 2-10 CADrawerController 中的常用函数

函数	说明
virtual bool initWithController(CAViewController * leftViewController, CAViewController * rightViewController,float division)	初始化 leftViewController：左边 rightViewController：右边 division：左边露出的尺寸
CAViewController * getLeftViewController()	获得左边的 ViewController
CAViewController * getRightViewController()	获得右边的 ViewController
void showLeftViewController(bool animated)	显示左边的 viewController animated：是否显示动画
void hideLeftViewController(bool animated)	隐藏左边的 viewController animated：是否显示动画
bool isShowLeftViewController()	左边是否在显示

它的创建和使用也非常简单，首先创建 LeftViewController 和 RightViewController，分别改变他们的背景色和文本内容。然后在 RootWindow 中引入它们的头文件，修改代码如下。

```
bool RootWindow::init()
{
    if (!CAWindow::init())
    {
        return false;
    }
```

```
//创建左边的 ViewController
LeftViewController* left = LeftViewController::create();
//创建右边的 ViewController
RightViewController* right = RightViewController::create();
CADrawerController* drawerController = new CADrawerController();
//最后参数是左边 ViewController 露出的尺寸
drawerController->initWithController(left, right, _px(300));
//必须加上背景,不然只能滑动一次
drawerController->setBackgroundView(CAView::create());
this->setRootViewController(drawerController);
//释放内存
drawerController->autorelease();
return true;
}
```

侧边抽屉式导航控制器滑动前和滑动后效果如图 2-8 所示。现在运行一下程序马上体验一下吧。

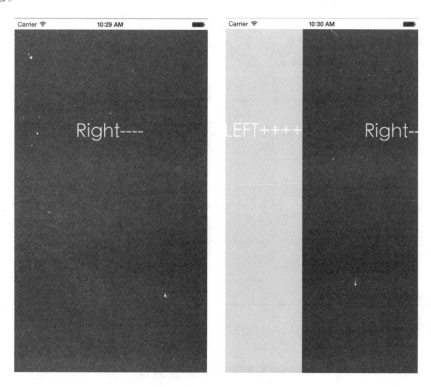

图 2-8　侧边抽屉式导航控制器滑动前和滑动后

第 3 章 CrossApp 核心控件与视图

核心控件指不同的 CAView 的子类，在使用的时候都可以方便地添加到视图控制器中，以满足不同 App 的 UI 需求。

3.1 文本 CALabel

CALabel 标签文字类，用于显示文本。

表 3-1 介绍了 CALabel 的常用方法

表 3-1 CALabel 常用方法

访问修饰符	属性名	说明
protected	Text	label 的文本内容
protected	fontName	label 的字体
protected	fontSize	label 的字体大小
protected	Color	label 的字体颜色
protected	VerticalTextAlignmet	文本的竖直对齐方式
protected	TextAlignment	文本的水平对齐方式
protected	NumberOfLine	label 的行数
访问修饰符	方法名	说明
public	Visit	更新 Label
Public	sizeToFit	设置自适应 label 宽度

官方文档详细地介绍了 CALabel 的使用方法，这里要注意以下几点问题。

（1）Label 的行数

其实 Label 行数是受 Label 的范围的影响，假如设置的 Frame 太小，超出 Frame 范围的行数是不被显示的。

例如，Frame 的范围只够显示 2 行，那么 NumberOfLine 设置的数值比 2 大也最多显示 2 行。

（2）Windows 开发中文字符

假如是在 Windows 下开发 CrossApp，如下代码：

```cpp
label->setText("中文");
```

直接设置中文字符，CALabel 的显示是不正常的。

可以设置转为 UTF8 格式，格式为

```cpp
label->setText(UTF8("伟大的矮人王索林·橡木盾"));
```

实例代码如下：

```cpp
void FirstViewController::viewDidLoad()
{
    CCRect winRect = this->getView()->getBounds();
    CAImageView* imageView =
        CAImageView::createWithImage(CAImage::create("HelloWorld.png"));
    imageView->setFrame(winRect);
    this->getView()->addSubview(imageView);
    //根据 Center 生成 CALabel
    CALabel* label =
        CALabel::createWithCenter(CCRect(winRect.size.width*0.5,
            winRect.size.height*0.5-270, winRect.size.width, 200));
    /*设置水平对齐样式
    CATextAlignmentLeft:左对齐
    CATextAlignmentCenter:居中
    CATextAlignmentRight:右对齐
    */
    label->setTextAlignment(CATextAlignmentCenter);
    /*设置竖直对齐样式
    CAVerticalTextAlignmentTop:顶部对齐
    CAVerticalTextAlignmentCenter:居中
    CAVerticalTextAlignmentBottom:底部对齐
    */
    label->setVerticalTextAlignmet(CAVerticalTextAlignmentCenter);
    //设置字体大小
    label->setFontSize(72 * CROSSAPP_ADPTATION_RATIO);
    //设置文本内容
    label->setText("Hello World!");
    //设置文本颜色为白色
    label->setColor(CAColor_white);
    //设置文本字体
    label->setFontName("fonts/arial.ttf");
    //将 label 添加渲染
    this->getView()->insertSubview(label, 1);
    //根据 Frame 创建一个 CALabel
    CALabel* firstLabel =
        CALabel::createWithFrame(CCRect(winRect.size.width*0.5,
            winRect.size.height*0.5, winRect.size.width * 0.5, 150));
    //设置行数为 3 行
```

```
        firstLabel->setNumberOfLine(3);
        //设置文本颜色为红色
        firstLabel->setColor(CAColor_red);
        //设置中文内容
        firstLabel->setText(UTF8("伟大的矮人王:橡木盾索林"));
        //添加到渲染
        this->getView()->addSubview(firstLabel);
    }
```

代码运行结果,如图 3-1 所示。

图 3-1　CALabel 运行效果

3.2　按钮 CAButton

按钮是应用中最常见的控件之一,CrossApp 中提供 CAButton 为按钮类,其主要为了接收用户的点击操作,从而触发特定的事件。

CAButton 默认为其设了三种类型,如图 3-2 所示。

(1) CAButtonTypeCuston:自定义类型(默认无背景,可以自定义图片或其他样式做背景)。

(2) CAButtonTypeSquareRect:直角类型。

(3) CAButtonTypeRoundedRect：圆角类型。

Hello

CAButtonTypeCuston    CAButtonTypeSquareRect    CAButtonTypeRoundedRect

图 3-2 CAButton 类型

CAButton 也为其设置了几种状态，列举如下。

(1) CAControlStateNormal：默认状态。
(2) CAControlStateHighlighted：高亮状态。
(3) CAControlStateDisabled：禁用状态。
(4) CAControlStateSelected：选中状态。
(5) CAControlStateAll：全部状态。
(6) CAButton 其主要属性如表 3-2 所示。

表 3-2 CAButton 成员函数

函 数	说 明
void setBackGroundViewForState(const CAControlState& controlState, CAView * var)	设置相应状态的背景图片
void setImage(CAImage * image)	设置相应状态的图片
void setTitleForState(const CAControlState& controlState, const std::string& var)	设置相应状态的文本内容
void setTitleColorForState(const CAControlState& controlState, const CAColor4B& var)	设置相应状态的文本颜色
void setTitleFontName(const std::string& var)	设置文本字体
void setControlState(CAControlState var)	设置按钮当前的状态
bool isTextTagEqual(const char * text)	判断当前按钮标签与指定文本是否相同
void interruptTouchState()	中断触摸事件，可以根据不同的条件中断触摸响应

CAButton 运行效果如图 3-3 所示，下面介绍实例代码。

1. FirstViewController.h

（1）源码

```
#ifndef __HelloCpp__ViewController__
#define __HelloCpp__ViewController__
#include <iostream>
#include "CrossApp.h"
USING_NS_CC;
class FirstViewController: public CAViewController
{
```

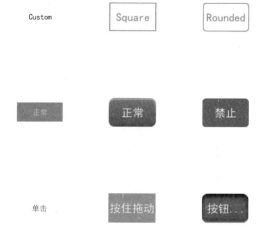

图 3-3 CAButton 运行效果

```
public:
    FirstViewController();
    virtual ~FirstViewController();
    //按钮样式
    void buttonType();
    //按钮状态及背景设置
    void buttonBackground();
    //按钮绑定监听事件
    void buttonTouchEvent();
    //监听回调函数
    void buttonCallback(CAControl* btn, CCPoint point);

protected:
    void viewDidLoad();
    void viewDidUnload();
};
#endif
```

(2) 代码解析

该类定义了几个按钮的设置函数和回调函数,代码如下:

```
//按钮样式
void buttonType();
//按钮状态及背景设置
void buttonBackground();
//按钮绑定监听事件
void buttonTouchEvent();
//监听回调函数
void buttonCallback(CAControl* btn, CCPoint point);
```

2. FirstViewController.cpp

代码如下:

```cpp
void FirstViewController::viewDidLoad()
{
    //CAButton 样式
    this->buttonType();
    //CAButton 背景及状态实例
    this->buttonBackground();
    //按钮绑定监听事件
    this->buttonTouchEvent();
}
//按钮样式
void FirstViewController::buttonType()
{
    //Custom 样式的 Button(无背景)
    CAButton * customBtn = CAButton::createWithCenter(CCRect(100, 100, 100, 40), CAButtonTypeCustom);
    //设置文本
    customBtn->setTitleForState(CAControlStateAll, "Coustom");
    //添加到渲染
    this->getView()->addSubview(customBtn);

    //Square 样式的 Button(方角)
    CAButton * squareBtn = CAButton::createWithCenter(CCRect(300, 100, 100, 40), CAButtonTypeSquareRect);
    //设置文本
    squareBtn->setTitleForState(CAControlStateAll, "Square");
    //添加到渲染
    this->getView()->addSubview(squareBtn);

    //Rounded 样式的 Button(圆角)
    CAButton * roundedBtn = CAButton::createWithCenter(CCRect(500, 100, 100, 40), CAButtonTypeRoundedRect);
    //设置文本
    roundedBtn->setTitleForState(CAControlStateAll, "Rounded");
    //添加到渲染
    this->getView()->addSubview(roundedBtn);
}

//按钮状态及背景设置
void FirstViewController::buttonBackground()
{
    //*****状态
    //创建一个 Custom 样式的 Button
    CAButton * defaultBtn = CAButton::create(CAButtonTypeCustom);
    //设置区域
```

```cpp
    defaultBtn->setCenter(CADipRect(100, 300, 100, 40));
    //设置默认状态显示的文本
    defaultBtn->setTitleForState(CAControlStateNormal, UTF8("正常"));
    //设置默认状态显示的文本颜色为白色
    defaultBtn->setTitleColorForState(CAControlStateNormal, CAColor_white);
    //设置选择状态显示的文本
    defaultBtn->setTitleForState(CAControlStateSelected, UTF8("选中"));
    //设置高亮状态显示的文本
    defaultBtn->setTitleForState(CAControlStateHighlighted, UTF8("高亮"));
    //设置默认状态显示的背景颜色为绿色
     defaultBtn->setBackGroundViewForState(CAControlStateNormal, CAView::createWithColor(CAColor_green));
    //设置高亮状态显示的背景颜色
     defaultBtn->setBackGroundViewForState(CAControlStateHighlighted, CAView::createWithColor(CAColor_yellow));
    this->getView()->addSubview(defaultBtn);

    //创建一个 Square 样式的 Button
    CAButton* squareRectBtn = CAButton::create(CAButtonTypeSquareRect);
    //设置可选中
    squareRectBtn->setAllowsSelected(true);
    squareRectBtn->setCenter(CADipRect(300, 300, 100, 40));
    //设置默认状态文本
    squareRectBtn->setTitleForState(CAControlStateNormal, UTF8("正常"));
    //设置默认字体文本颜色为白色
    squareRectBtn->setTitleColorForState(CAControlStateNormal, CAColor_white);
    //设置选中状态文本
    squareRectBtn->setTitleForState(CAControlStateSelected, UTF8("选中"));
    //设置高亮状态文本
    squareRectBtn->setTitleForState(CAControlStateHighlighted, UTF8("高亮"));
    //设置默认状态背景图片
     squareRectBtn->setBackGroundViewForState(CAControlStateNormal, CAScale9ImageView::createWithImage(CAImage::create("source_material/btn_rounded3D_normal.png")));
    //设置高亮状态背景图片
     squareRectBtn->setBackGroundViewForState(CAControlStateHighlighted, CAScale9ImageView::createWithImage(CAImage::create("source_material/ex4.png")));
    //设置选择状态背景图片
     squareRectBtn->setBackGroundViewForState(CAControlStateSelected, CAScale9ImageView::createWithImage(CAImage::create("source_material/btn_rounded3D_selected.png")));
    this->getView()->addSubview(squareRectBtn);

    //创建一个 Rounded 样式的 Button
    CAButton* roundedRectBtn = CAButton::create(CAButtonTypeRoundedRect);
    roundedRectBtn->setCenter(CADipRect(500, 300, 100, 40));
    //设置禁用状态背景图片
    roundedRectBtn->setBackGroundViewForState(CAControlStateDisabled, CAScale9ImageView::createWithImage(CAImage::create("source_material/btn_rounded_selected.png")));
```

```cpp
    //设置Button当前状态为禁用
    roundedRectBtn->setControlState(CAControlStateDisabled);
    //设置所有状态显示文本
    roundedRectBtn->setTitleForState(CAControlStateAll, UTF8("禁止"));
    //添加到绘制
    this->getView()->addSubview(roundedRectBtn);


}

//按钮绑定监听事件
void FirstViewController::buttonTouchEvent()
{
    //创建一个无背景的Button
    CAButton* btnOne = CAButton::create(CAButtonTypeCustom);
    //设置Button大小
    btnOne->setCenter(CADipRect(100,500,100,40));
    //设置tag值
    btnOne->setTag(1);
    //设置全部状态的文本
    btnOne->setTitleForState(CAControlStateAll, UTF8("单击"));
    //设置默认文本颜色
    btnOne->setTitleColorForState(CAControlStateNormal, CAColor_green);
    //设置默认背景图片
    btnOne->setBackGroundViewForState(CAControlStateNormal, CAScale9ImageView::createWithImage
(CAImage::create("source_material/round1.png")));
    //设置高亮状态时背景图片
    btnOne->setBackGroundViewForState(CAControlStateHighlighted, CAScale9ImageView::createWithImage
(CAImage::create("source_material/round2.png")));
    //绑定按钮被按下时的回调函数
    btnOne->addTarget(this, CAControl_selector(FirstViewController::buttonCallback),
CAControlEventTouchDown);
    this->getView()->addSubview(btnOne);

    //创建一个直角Button
    CAButton* btnTwo = CAButton::create(CAButtonTypeSquareRect);
    //设置Button大小
    btnTwo->setCenter(CADipRect(300, 500, 100, 40));
    //设置tag值
    btnTwo->setTag(2);
    //设置全部状态的文本
    btnTwo->setTitleForState(CAControlStateAll, UTF8("按住拖动"));
    //设置默认文本颜色
    btnTwo->setTitleColorForState(CAControlStateNormal, CAColor_white);
    //设置默认背景图片
    btnTwo->setBackGroundViewForState(CAControlStateNormal, CAScale9ImageView::createWithImage
(CAImage::create("source_material/btn_square_highlighted.png")));
```

```cpp
    //设置高亮状态时背景图片
    btnTwo->setBackGroundViewForState(CAControlStateHighlighted, CAScale9ImageView::createWithImage
(CAImage::create("source_material/btn_square_selected.png")));
    //绑定选中按钮后移动的回调函数
    btnTwo->addTarget(this, CAControl_selector(FirstViewController::buttonCallback),
CAControlEventTouchMoved);
    this->getView()->addSubview(btnTwo);
    
    CAButton* btnThree = CAButton::create(CAButtonTypeRoundedRect);
    //设置Button大小
    btnThree->setCenter(CADipRect(500, 500, 100, 40));
    //设置tag值
    btnThree->setTag(3);
    //设置全部状态的文本
    btnThree->setTitleForState(CAControlStateAll, UTF8("按钮内单击"));
    //设置默认文本颜色
    btnThree->setTitleColorForState(CAControlStateNormal, CAColor_white);
    //设置默认背景图片
    btnThree->setBackGroundViewForState(CAControlStateNormal, CAScale9ImageView::createWithImage
(CAImage::create("source_material/btn_rounded3D_highlighted.png")));
    //设置高亮状态时背景图片
    btnThree->setBackGroundViewForState(CAControlStateHighlighted, CAScale9ImageView::
createWithImage(CAImage::create("source_material/btn_rounded3D_selected.png")));
    //绑定按钮抬起时的回调函数
    btnThree->addTarget(this, CAControl_selector(FirstViewController::buttonCallback),
CAControlEventTouchUpInSide);
    this->getView()->addSubview(btnThree);
    
}
void FirstViewController::buttonCallback(CAControl* btn, CCPoint point)
{
    //将btn转成CAButton类型
    CAButton* button = (CAButton*)btn;
    //获取tag值
    int tag = btn->getTag();
    //打印
    CCLog("ButtonTag:%d, x:%0.2f,y:%.02f",tag, point.x, point.y);
}
```

3.3 图片 CAImageView

CAImageVies是图片空间,用来将图片显示在屏幕上。CAImageView继承CAView,通过CAImage加载纹理创建。

CAImage其主要函数如表3-3所示。

表 3-3 CAImage 主要函数

函数	说明
void setImage(CAImage * image)	imageView 所显示的图像
virtual void setImageViewScaleType (const CAImageViewScaleType& var)	imageView 的缩放模式
static CAImageView * createWithImage (CAImage * image)	创建一个 imageView，传递一个 CAImage 对象作为显示内容
virtual void setImageAsyncWithFile(const char * fileName)	创建一个 imageView，传递一个 CAImage 对象作为显示内容

实例代码如下：

```
void FirstViewController::viewDidLoad()
{
    //获得屏幕大小
    CCRect winRect = this->getView()->getBounds();
    //通过 CAImage 创建一个 CAImgeView
    CAImageView * imageView =
        CAImageView::createWithImage(CAImage::create("HelloWorld.png"));
    //设置显示区域
    imageView->setFrame(winRect);
    //添加渲染
    this->getView()->addSubview(imageView);
    //创建一个 CAImage
    CAImage * image = CAImage::create("source_material/btn_right_blue.png");
    //通过 CAImage 创建一个 CAImageView
    CAImageView * firstIV = CAImageView::createWithImage(image);
    //设置显示区域
    firstIV->setFrame(CCRect(100, 100,
        image->getContentSize().width, image->getContentSize().height));
    //添加渲染
    this->getView()->addSubview(firstIV);
    //通过显示区域创建 CAImageView
    CAImageView * secondIV = CAImageView::createWithFrame(CCRect(200,100,65,65));
    //异步加载图片
    secondIV->setImageAsyncWithFile("source_material/btn_left_blue.png");
    //添加渲染
    this->getView()->addSubview(secondIV);
}
```

3.4 九宫格图片 CAScale9ImageView

CAScale9ImageView 是 CrossApp 提供的一种九宫格拉伸图片的解决方案,首先来了解一下什么是九宫格图片拉伸。

在 App 的设计过程中,为了适配不同的手机分辨率,图片大小需要拉伸或者压缩,这样就出现了可以任意调整大小的拉伸样式,如图 3-4 所示。

图 3-4 九宫格缩放的图片

CAScale9ImageView 的实现非常巧妙,通过将原纹理资源切割成 9 部分(这也是称作九宫图的原因),根据需要的尺寸,完成以下三个步骤:

(1) 保持 4 个角部分不变形;
(2) 单向拉伸 4 条边(即在 4 个角两两之间的边,比如上边只做横向拉伸);
(3) 双向拉伸中间部分(即九宫图的中间部分,横向、纵向同时拉伸,拉伸比例不一定相同)。

九宫格缩放原理如图 3-5 所示。

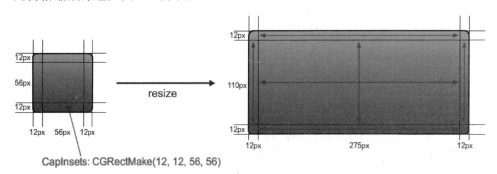

图 3-5 九宫格缩放原理

CAScale9ImageView 的实例代码如下:

```
void FirstViewController::viewDidLoad()
{
```

```cpp
//创建
CAScale9ImageView * first9IV = 
    CAScale9ImageView::createWithImage(
        CAImage::create("source_material/btn_rounded_normal.png"));
//设置非拉伸区域为(5,5,20,20)
first9IV->setCapInsets(CCRect(5,5,20,20));
//设置显示大小(拉伸后的大小)
first9IV->setFrame(CCRect(100, 100, 50, 140));
//添加渲染
this->getView()->addSubview(first9IV);
//创建
CAScale9ImageView * second9IV = 
    CAScale9ImageView::createWithImage(
        CAImage::create("source_material/btn_rounded_highlighted.png"));
//设置非拉伸区域与顶部的距离
second9IV->setInsetTop(3.0f);
//设置非拉伸区域与底部的距离
second9IV->setInsetBottom(3.0f);
//设置非拉伸区域与右边的距离
second9IV->setInsetRight(3.0f);
//设置非拉伸区域与左边的距离
second9IV->setInsetLeft(3.0f);
//设置显示区域(拉伸后的大小)
second9IV->setFrame(CCRect(400, 200, 150, 40));
//添加渲染
this->getView()->addSubview(second9IV);
//创建
CALabel * image9Label = CALabel::createWithFrame(CCRect(400, 200, 150, 40));
image9Label->setText(UTF8("使用九图"));
image9Label->setTextAlignment(CATextAlignmentCenter);
image9Label->setVerticalTextAlignmet(CAVerticalTextAlignmentCenter);
this->getView()->addSubview(image9Label);
//不使用九宫格拉伸 设置大小同上
CAImageView * imageView = 
    CAImageView::createWithFrame(CCRect(400, 300, 150, 40));
//设置显示图片
imageView->setImage(
        CAImage::create("source_material/btn_rounded_highlighted.png"));
//添加渲染
this->getView()->addSubview(imageView);
//创建
CALabel * imageLabel = CALabel::createWithFrame(CCRect(400, 300, 150, 40));
imageLabel->setText(UTF8("不使用九图"));
imageLabel->setTextAlignment(CATextAlignmentCenter);
imageLabel->setVerticalTextAlignmet(CAVerticalTextAlignmentCenter);
this->getView()->addSubview(imageLabel);
}
```

3.5 单行输入框 CATextField

CATextField 是单行输入框控件，主要接收用户的文本输入，多用于用户名、密码和聊天输入等。

CATextField 的常用函数如表 3-4 所示。

表 3-4 CATextField 常用函数

函　　数	说　　明
void setBackGroundImage(CAImage * image)	输入框背景图片
void setPlaceHolder(const std::string &var)	默认的提示文本
void setCursorColor(const CAColor4B &var)	设置光标的颜色
void setInputType(eKeyBoardInputType type)	输入的类型
inline void setKeyboardType (eKeyBoardType type)	键盘的类型(真机或模拟器上有效)
inline void setKeyboardReturnType (eKeyBoardReturnType type)	设置确认键的类型(真机或模拟器上有效)

在 CATextField 接收用户输入文本时，有时候希望获得用户的操作行为，比如 CATextField 获得焦点，CATextField 失去焦点，用户输入字符和用户删除字符等，这样可以对用户的操作进行逻辑处理，比如限制输入内容和输入字符长度等。那么如何才能监听到 CATextField 的改变呢？我们需要了解一下 CATextFieldDelegate，它主要使用四个函数，分别是

```
//获得焦点
virtual bool onTextFieldAttachWithIME(CATextField * sender);
//失去焦点
virtual bool onTextFieldDetachWithIME(CATextField * sender);
//输入文本
virtual bool onTextFieldInsertText(CATextField * sender, const char * text, int nLen);
//删除文本
virtual bool onTextFieldDeleteBackward(CATextField * sender,const char * delText, int nLen)
```

假如想在 FirstViewController 中监听 CATextField，那么我们需要使 FirstViewController 继承 CATextFieldDelegate 并重写这些函数。

下面以图 3-6 所示的登录界面为例，介绍 CATextField 的使用方法。

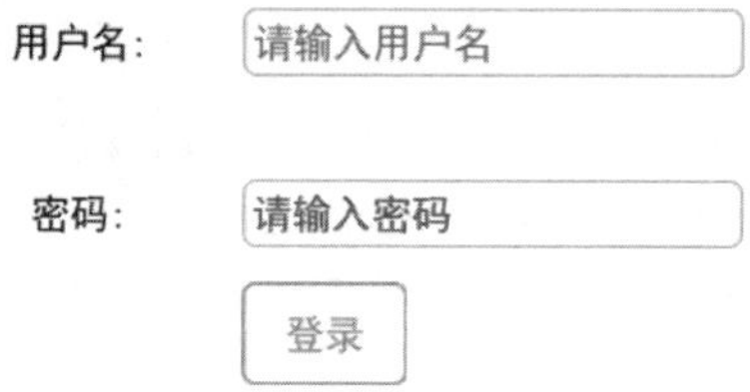

图 3-6 使用 CATextField 实现用户登录界面

1. FirstViewController.h

（1）源码

```
#include <iostream>
#include "CrossApp.h"
USING_NS_CC;
class FirstViewController : public CAViewController, public CATextFieldDelegate
{
public:
    FirstViewController();
    virtual ~FirstViewController();
    //获得焦点
    virtual bool onTextFieldAttachWithIME(CATextField * sender);
    //失去焦点
    virtual bool onTextFieldDetachWithIME(CATextField * sender);
    //输入文本
    virtual bool onTextFieldInsertText(CATextField * sender, const char * text, int nLen);
    //删除文本
    virtual bool onTextFieldDeleteBackward(CATextField * sender, const char * delText, int nLen);
    //登录
    void login(CAControl * control,CCPoint point);
protected:

    void viewDidLoad();
    void viewDidUnload();
};
```

（2）代码解析

以上 FirstViewController 继承 CATextFieldDelegate，实现了本类是一个 CATextField 的委托类，并在成员中覆盖了相应的回调函数。

2. FirstViewController.cpp

（1）源码

```
#include "FirstViewController.h"
FirstViewController::FirstViewController()
{
}
FirstViewController::~FirstViewController()
{
}
void FirstViewController::viewDidLoad()
{
    //用户名文本
    CALabel * nameLabel = CALabel::createWithFrame(CCRect(50, 100, 100, 40));
    nameLabel->setText(UTF8("用户名:"));
```

```cpp
nameLabel->setTextAlignment(CATextAlignmentCenter);
nameLabel->setVerticalTextAlignmet(CAVerticalTextAlignmentCenter);
this->getView()->addSubview(nameLabel);
//密码文本
CALabel* passwordLabel = CALabel::createWithFrame(CCRect(50, 200, 100, 40));
passwordLabel->setText(UTF8("密码:"));
passwordLabel->setTextAlignment(CATextAlignmentCenter);
passwordLabel->setVerticalTextAlignmet(CAVerticalTextAlignmentCenter);
this->getView()->addSubview(passwordLabel);
//创建
CATextField* nameTF = CATextField::createWithFrame(CCRect(200, 100, 300, 40));
//设置tag
nameTF->setTag(1);
//设置提示文本
nameTF->setPlaceHolder(UTF8("请输入用户名"));
//设置光标颜色
nameTF->setCursorColor(CAColor_orange);
/*设置键盘类型(真机或模拟器上有效)
KEY_BOARD_TYPE_NORMAL:普通键盘
KEY_BOARD_TYPE_NUMBER:数字键盘
KEY_BOARD_TYPE_ALPHABET:字母键盘
*/
nameTF->setKeyboardType(eKeyBoardType::KEY_BOARD_TYPE_ALPHABET);
/*设置确认键的类型(真机或模拟器上有效)
KEY_BOARD_RETURN_DONE:完成
KEY_BOARD_RETURN_SEARCH:搜索
KEY_BOARD_RETURN_SEND:发送
*/
nameTF->setKeyboardReturnType(eKeyBoardReturnType::KEY_BOARD_RETURN_DONE);
//绑定代理(设置代理才能被监听状态)
nameTF->setDelegate(this);
//添加渲染
this->getView()->addSubview(nameTF);
CATextField* password = CATextField::createWithFrame(CCRect(200,200,300,40));
//设置tag
password->setTag(2);
//设置提示文本
password->setPlaceHolder(UTF8("请输入密码"));
//设置提示文本颜色
password->setSpaceHolderColor(CAColor_red);
//设置输入样式为密码格式
password->setInputType(eKeyBoardInputType::KEY_BOARD_INPUT_PASSWORD);
//添加渲染
this->getView()->addSubview(password);
//登录按钮
```

```cpp
    CAButton* loginBtn = CAButton::createWithFrame(
        CCRect(200, 260, 100, 40), CAButtonType::CAButtonTypeRoundedRect);
    loginBtn->setTitleForState(CAControlStateAll, UTF8("登录"));
    loginBtn->addTarget(this,
        CAControl_selector(FirstViewController::login), CAControlEventTouchDown);
    this->getView()->addSubview(loginBtn);
}
void FirstViewController::viewDidUnload()
{
    // Release any retained subviews of the main view.
    // e.g. self.myOutlet = nil;
}
bool FirstViewController::onTextFieldAttachWithIME(CATextField * sender)
{
    //获得焦点
    CCLog("onTextFieldAttachWithIME --->");
    return false;
}
bool FirstViewController::onTextFieldDetachWithIME(CATextField * sender)
{
    //失去焦点
    CCLog("onTextFieldDetachWithIME --->");
    return false;
}
bool FirstViewController::onTextFieldInsertText(CATextField * sender, const char * text, int nLen)
{
    //输入时调用
    CCLog("onTextFieldInsertText --->Text:%s,Len:%d", text, nLen);
    return false;
}
bool FirstViewController::onTextFieldDeleteBackward(CATextField * sender, const char * delText, int nLen)
{
    //删除字符时调用
    CCLog("onTextFieldDeleteBackward --->Text:%s,Len:%d", delText, nLen);
    return false;
}
//登录
void FirstViewController::login(CAControl* control, CCPoint point)
{
    //根据tag值获得nameTF和passwordTF
    CATextField* nameTF = (CATextField*)this->getView()->getSubviewByTag(1);
    CATextField* passwordTF = (CATextField*)this->getView()->getSubviewByTag(2);
    //获得输入框的内容
```

```
            string name = nameTF->getText();
            string password = passwordTF->getText();
            //如果用户名为"9miao" 密码为"123456" 则打印ok,否则打印error
            if (strcmp(name.c_str(), "9miao") == 0 && strcmp(password.c_str(), "123456") == 0)
            {
                CCLog("OK");
            }
            else
            {
                CCLog("ERROR");
            }
    }
```

(2)代码解析

通过对回调函数实现,定义了一个最简单的登录页面以及当用户输入之后的处理逻辑,通过以下方法实现了对按钮按下的回调定义:

```
loginBtn->addTarget(this,
         CAControl_selector(FirstViewController::login), CAControlEventTouchDown);
```

通过以下方法实现了对输入控件的回调处理:

```
nameTF->setDelegate(this);
```

3.6 多行输入框 CATextView

上一节讲到 CATextField 用于单行输入,假如要输入的文字比较多,并且需要在一定区域里展示出来,就可以选择多行输入框 CATextView。

CATextView 是多行输入框,其主要用法和 CATextField 相似,表 3-5 是 CATextView 的主要函数。

表 3-5 CATextView 主要函数

函　　数	说　　明
void setBackGroundImage(CAImage * image)	输入框背景图片
void setPlaceHolder(const std::string &var)	默认的提示文本
void setCursorColor(const CAColor4B &var)	设置光标的颜色
void setInputType(eKeyBoardInputType type)	输入的类型
inline void setKeyboardType (eKeyBoardType type)	键盘的类型(真机或模拟器上有效)
inline void setKeyboardReturnType (eKeyBoardReturnType type)	设置确认键的类型(真机或模拟器上有效)

图 3-7 展示了使用 CATextVeiw 实现多行文字输入的界面，下面介绍实例代码。

图 3-7 使用 CATextVeiw 实现多行文字输入

1. FirstViewController.h

（1）源码

```
#include <iostream>
#include "CrossApp.h"
USING_NS_CC;
class FirstViewController: public CAViewController ,public CATextViewDelegate
{
public:
    FirstViewController();
    
    virtual ~FirstViewController();
    //获得焦点
    virtual bool onTextFieldAttachWithIME(CATextField * sender);
    //失去焦点
    virtual bool onTextFieldDetachWithIME(CATextField * sender);
    //输入文本
    virtual bool onTextFieldInsertText(CATextField * sender, const char * text, int nLen);
    //删除文本
    virtual bool onTextFieldDeleteBackward(CATextField * sender, const char * delText, int nLen);
protected:
    void viewDidLoad();
    void viewDidUnload();
};
```

(2) 代码解析

在 CATextView 也有相应的回调，需要委托类继承 CATextViewDelegate。

2. FirstViewController.cpp

代码如下：

```cpp
void FirstViewController::viewDidLoad()
{
    CCRect winRect = this->getView()->getBounds();
    CAImageView* imageView =
        CAImageView::createWithImage(CAImage::create("HelloWorld.png"));
    imageView->setFrame(winRect);
    this->getView()->addSubview(imageView);
    CATextView* tw = CATextView::createWithFrame(CCRect(100, 100, 300, 400));
    //设置提示文本
    tw->setPlaceHolder(UTF8("多行输入框"));
    //设置文本字体大小
    tw->setFontSize(20);
    //设置文本是否折行
    tw->setWordWrap(true);
    //设置行距
    tw->setLineSpacing(10);
    /*设置键盘类型(真机或模拟器上有效)
    KEY_BOARD_TYPE_NORMAL:普通键盘
    KEY_BOARD_TYPE_NUMBER:数字键盘
    KEY_BOARD_TYPE_ALPHABET:字母键盘
    */
    tw->setKeyboardType(eKeyBoardType::KEY_BOARD_TYPE_ALPHABET);
    /*设置确认键的类型(真机或模拟器上有效)
    KEY_BOARD_RETURN_DONE:完成
    KEY_BOARD_RETURN_SEARCH:搜索
    KEY_BOARD_RETURN_SEND:发送
    */
    tw->setKeyboardReturnType(eKeyBoardReturnType::KEY_BOARD_RETURN_DONE);
    //事件代理参考 CATextField
    tw->setTextViewDelegate(this);
    //加入渲染
    this->getView()->addSubview(tw);
}

void FirstViewController::viewDidUnload()
{
    // e.g. self.myOutlet = nil;
}
bool FirstViewController::onTextFieldAttachWithIME(CATextField* sender)
{
    //获得焦点
```

```
        CCLog("onTextFieldAttachWithIME--->");
        return false;
}
bool FirstViewController::onTextFieldDetachWithIME(CATextField * sender)
{
        //失去焦点
        CCLog("onTextFieldDetachWithIME--->");
        return false;
}

bool FirstViewController::onTextFieldInsertText(CATextField * sender, const char * text,
int nLen)
{
        //输入时调用
        CCLog("onTextFieldInsertText--->Text:%s,Len:%d", text, nLen);
        return false;
}
bool FirstViewController::onTextFieldDeleteBackward(CATextField * sender, const char *
delText, int nLen)
{
        //删除字符时调用
        CCLog("onTextFieldDeleteBackward--->Text:%s,Len:%d", delText, nLen);
        return false;
}
```

3.7 开关 CASwitch

CASwitch 控件是开关控件,可以实现类型开关的效果,如图 3-8 所示。

图 3-8 CASwich 控件

CASwitch 常用的函数如表 3-6 所示。

表 3-6 CASwitch 控件常用函数

函 数 名	说 明
void setOnImage(CAImage * onImage)	switch 开状态时的图片
void setOffImage(CAImge * offImage)	switch 关状态时的图片
void setThumTintImage(CAImage * thumbTintImage)	switch 背景图片
bool isOn()	switch 是否处于开状态

CASwitch 使用起来也非常简单，实例代码如下：

1. FirstViewController.cpp

（1）添加以下代码。

```cpp
void FirstViewController::viewDidLoad()
{
    // Do any additional setup after loading the view from its nib.
    CCSize size = this->getView()->getBounds().size;
    //创建
    CASwitch* defaultSwitch = CASwitch::createWithCenter(
        CADipRect(size.width * 0.5, size.height * 0.2, size.width * 0.3, 20));
    //设置tag
    defaultSwitch->setTag(100);
    //设置监听函数
    defaultSwitch->addTarget(this, CAControl_selector(FirstViewController::callback));
    //添加绘制
    this->getView()->addSubview(defaultSwitch);
    //创建
    CASwitch* customSwitch = CASwitch::createWithCenter(
        CADipRect(size.width * 0.5, size.height * 0.4, size.width * 0.3, 20));
    //设置tag
    customSwitch->setTag(101);
    //设置开启时图片
    customSwitch->setOnImage(
        CAImage::create("source_material/btn_rounded_highlighted.png"));
    //设置关闭时图片
    customSwitch->setOffImage(
        CAImage::create("source_material/btn_rounded_normal.png"));
    //设置中间图片
    customSwitch->setThumbTintImage(
        CAImage::create("source_material/btn_rounded3D_selected.png"));
    //设置监听函数
    customSwitch->addTarget(this, CAControl_selector(FirstViewController::callback));
    //添加绘制
    this->getView()->addSubview(customSwitch);
}
//监听函数
void FirstViewController::callback(CAControl* control, CCPoint point)
{
    CCLog("callback");
    //强转类型
    CASwitch* caSwtich = (CASwitch*)control;
    //获得tag
    CCLog("Tag:%d", caSwtich->getTag());
    //获得状态
```

```
    if (!caSwtich->isOn())
    {
        CCLog("OFF");
    }
    else
    {
        CCLog("ON");
    }
}
```

(2)代码解析

处理开关控件首先需要在成员中添加一个回调函数,格式如下:

`void FirstViewController::callback(CAControl * control, CCPoint point)`

函数名称可以随意,但传递的参数要按照如上形式,然后在定义完开关控件之后通过以下代码定义开关的事件响应函数:

`customSwitch->addTarget(this, CAControl_selector(FirstViewController::callback));`

这样就在屏幕上看到圆形开关和方形开关了,当用户改变开关状态,callback 函数会被回调。

3.8 提示框 CAAlertView

CAAlertView 是提示框控件,如果提示框内的按钮个数不超过三个,则横向排列按钮,如果按钮个数超过三个,则纵向排列按钮。

表 3-7 是 CAAlertView 常用的一些函数。

表 3-7 CAAlertView 常用函数

函 数	说 明
void setMessageFontName(std::string &var)	提示信息的字体
void setTitle(std::string var, CAColor4B col)	提示框的标题
void setAlertMessage(std::string var, CAColor4B col)	提示框的提示信息
CAAlertView * createWithText(const char * pszTitle, const char * pszAlertMsg, const char * pszBtnText, …)	创建 CAAlertView
void show()	显示提示框
void addButton(const std::string& btnText, CAColor4B col = ccc4(3, 100, 255, 255), CAImage * pNormalImage = NULL, CAImage * pHighlightedImage = NULL);	添加一个按钮到 CAAlertView
void setTarget(CAObject * target, SEL_CAAlertBtnEvent selector);	添加监听

要想获取用户触摸 CAAlertView，需要首先在视图控制器中声明函数：

```cpp
//提示框的回调函数
void alertViewCallback(int btnIndex);
```

1. FistViewContrdler.cpp

（1）添加以下代码

```cpp
void FirstViewController::viewDidLoad()
{
    //获取屏幕宽度
    CCSize size = this->getView()->getBounds().size;
    //设置背景颜色为黑色
    this->getView()->setColor(CAColor_black);
    //创建 Button
    CAButton* imageBtn = CAButton::createWithCenter(
        CADipRect(size.width * 0.5, 500, 200, 50), CAButtonTypeSquareRect);
    //设置 Buttion 文本
    imageBtn->setTitleForState(CAControlStateAll, "Click");
    //设置 tag 值
    imageBtn->setTag(1);
    //设置按钮监听
    imageBtn->addTarget(this,
        CAControl_selector(FirstViewController::respondTouch),
        CAControlEventTouchUpInSide);
    //添加到屏幕
    this->getView()->addSubview(imageBtn);
}
void FirstViewController::respondTouch(CAControl* btn, CCPoint point)
{
    //获得屏幕大小
    CCSize size = this->getView()->getBounds().size;
    //创建 CAAlerView 并设置显示文本和 green 按钮和 yellow 按钮
    CAAlertView* alertView = CAAlertView::createWithText("ButtonImage",
        UTF8("点击替换按钮颜色"), "green", "yellow", NULL);
    //获得 0~1 之间的随机数
    float randNum = CCRANDOM_0_1();
    if (randNum > 0.333f)
    {
        //添加按钮设置文本为 orange
        alertView->addButton("orange");
    }
    if (randNum > 0.666f)
    {
        //添加按钮并设置文本为 blue
        alertView->addButton("blue");
    }
```

```cpp
    //显示弹窗(如果不调用,弹窗不显示)
    alertView->show();
    //设置弹窗按钮的回调
    alertView->setTarget(this, CAAlertView_selector(FirstViewController::alertViewCallback));
}
void FirstViewController::alertViewCallback(int btnIndex)
{
    //根据tag获得imageBtn对象
    CAButton* imageBtn = (CAButton*)this->getView()->getSubviewByTag(1);
    //根据CAAlertView上按钮的index判断响应的逻辑
    if (btnIndex == 0)
    {
        //设置imageBtn背景色为green
        imageBtn->setBackGroundViewForState(CAControlStateNormal,
            CAView::createWithColor(CAColor_green));
    }
    else if (btnIndex == 1)
    {
        //设置imageBtn背景色为yellow
        imageBtn->setBackGroundViewForState(CAControlStateNormal,
            CAView::createWithColor(CAColor_yellow));
    }
    else if (btnIndex == 2)
    {
        //设置imageBtn背景色为orange
        imageBtn->setBackGroundViewForState(CAControlStateNormal,
            CAView::createWithColor(CAColor_orange));
    }
    else
    {
        //设置imageBtn背景色为blue
        imageBtn->setBackGroundViewForState(CAControlStateNormal,
            CAView::createWithColor(CAColor_blue));
    }
}
```

(2) 代码解析

在viewDidLoad()函数中添加了一个Button,并为这个Button绑定监听,当按下这个Button时调用:respondTouch(CAControl* btn, CCPoint point),创建一个提示框,根据一个随机数来判断提示框中的按钮格式是两个还是三个,以下代码为CAAlertView绑定监听函数:

```cpp
alertView->setTarget(this, CAAlertView_selector(FirstViewController::alertViewCallback));
```

3.9 进度条 CAProgress

CAProgress 是进度条控件，主要用于显示任务进度，如图 3-9 所示。CAProgress 主要函数如表 3-8 所示。

CAProgress 是一个很简单的控件，其使用方式也比较便捷，只需要在相应的逻辑里对齐值进行增减即可，下面实例演示了在 CrossApp 的定时器中，每间隔一段时间增减 CAProgress 的值，当 CAProgress 值超过最大时设置为 0。

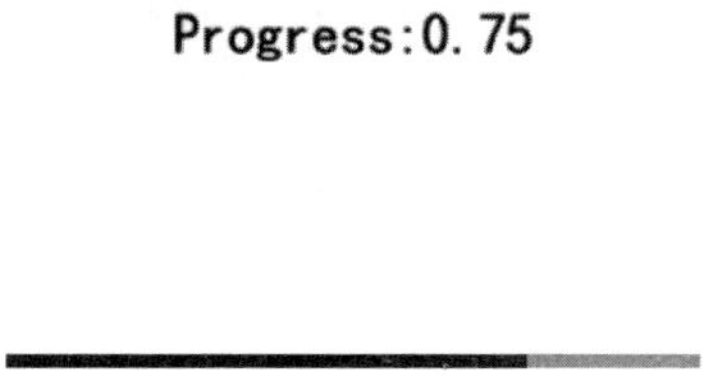

Progress:0.75

图 3-9 进度条控件使用

表 3-8 CAProgress 主要函数

函　　数	说　　明
void setProgress(float progress)	设置进度的值范围 0~1 之间的浮点数
float getProgress()	获得当期的进度值
voidsetProgressStyle(const CrossApp::CAProgressStyle&.var)	设置进度条的样式

1. FirstViewController.h

（1）源码

```cpp
#ifndef __HelloCpp__ViewController__
#define __HelloCpp__ViewController__
#include <iostream>
#include "CrossApp.h"
USING_NS_CC;
class FirstViewController: public CAViewController
{
public:
    FirstViewController();
    virtual ~FirstViewController();
    //定时器函数
    void updateProgressValue(float dt);
protected:
    void viewDidLoad();
    void viewDidUnload();
};
#endif
```

（2）代码解析

以上在视图控制器中定义一个定时器函数，该函数为了实现定时回调，修改进度条进度功能，其结构如下：

```cpp
//定时器函数
void updateProgressValue(float dt);
```

2. FirstViewController.cpp

（1）添加以下代码。

```cpp
void FirstViewController::viewDidLoad()
{
    // Do any additional setup after loading the view from its nib.
    CCSize size = this->getView()->getBounds().size;
    CAProgress* progress = CAProgress::create();
    //设置进度条样式
    progress->setProgressStyle(CAProgressStyle::CAProgressStyleBar);
    //设置显示区域
    progress->setCenter(CCRect(size.width * 0.5, 200, 300, 60));
    //设置进度值(0--1)之间的 float
    progress->setProgress(0.5f);
    //设置进度的颜色
    progress->setProgressTintColor(CAColor_orange);
    //设置进度的图片
    progress->setProgressTintImage(
        CAImage::create("source_material/btn_rounded_highlighted.png"));
    //设置背景的颜色
    progress->setProgresstrackColor(CAColor_yellow);
    //设置背景的图片
    progress->setProgressTrackImage(
        CAImage::create("source_material/btn_rounded3D_selected.png"));
    //设置 tag 值
    progress->setTag(1);
    //添加到屏幕
    this->getView()->addSubview(progress);
    //创建 Label 用于显示 progress 的值
    CALabel* label = CALabel::createWithCenter(CCRect(size.width * 0.5, 100, 200, 100));
    //水平居中
    label->setTextAlignment(CATextAlignmentCenter);
    //显示 progress 的值
    label->setText(CCString::createWithFormat("Progress:%.02f", progress->getProgress())->getCString());
    //设置 tag 值
    label->setTag(2);
    //添加到屏幕
    this->getView()->addSubview(label);
```

```cpp
    //启动定时器,间隔0.05s 调用
    CAScheduler::schedule(schedule_selector(FirstViewController::updateProgressValue), this, 0.05, false);
}
//定时器函数
void FirstViewController::updateProgressValue(float dt)
{
    //根据 tag 获得 progress 对象
    CAProgress * progress = (CAProgress *) this->getView()->getSubviewByTag(1);
    //获得 progress 的值
    float value = progress->getProgress();
    if (value < 1.0f)
    {
        value = value + 0.01;
    }
    else
    {
        value = 0;
    }
    //赋值
    progress->setProgress(value);
    //根据 tag 获得 label
    CALabel * label = (CALabel *)this->getView()->getSubviewByTag(2);
    //显示 value 值
    label->setText(CCString::createWithFormat("Progress:%.02f", value)->getCString());
}
```

(2) 代码解析

通过如下代码可以定时调用定时器函数 updateProgressValue：

CAScheduler::schedule(schedule_selector(函数 updateProgressValue), this, 0.05, false);

然后在该函数中可以完成逻辑编程，在实际项目中也会经常使用定时器。

3.10 滚动条 CASlider

CASlider 是滚动条控件，主要作用是方便数值调节，如音量大小控制、缩放视图等操作。如图 3-10 所示。

图 3-10　滚动条控件

表 3-9 是 CASlider 的常用函数。

表 3-9　CASlider 常用函数

函　　数	说　　明
void setValue(float value)	设置当前值
void setMinValue(float minValue)	设置最小值
void setMaxValue(float maxValue)	设置最大值
void setTrackHeight(float trackHeight)	设置高度
void setMinTrackTintImage(CAImage * image)	设置前景图片(已经划过的部分)
void setMaxTrackTintImage(CAImage * image)	设置后景图片(未划过的部分)
void setThumbTintImage(CAImage * image)	设置滚动块的图片

通过设置 CASlider 的最大值和最小值来确定其范围,可以通过设置图片改变滚动条的样式。下面是利用 CASlider 来控制一张图片的缩放大小的示例,说明 CASlider 的使用方法。

首先在 FirstViewController.h 中添加一个监听函数,用于监听 CASlider 的值的变化。

(1) 源码

```
//监听函数
void zoomViewBySliderValue(CAControl * control, CCPoint point)
```

然后在 FirstViewController.cpp 中去实现 CASlider 来控制 CAImageView 的缩放变化。

```
void FirstViewController::viewDidLoad()
{
    //获得屏幕大小
    CCSize size = this->getView()->getBounds().size;
    //创建 CAImageView
    CAImageView * imageView =
        CAImageView::createWithImage(CAImage::create("HelloWorld.png"));
    imageView->setCenter(CADipRect(size.width * 0.5, size.height * 0.5, 800, 1200));
    //设置 tag
    imageView->setTag(1);
    //添加
    this->getView()->addSubview(imageView);
    //创建 CASlider
    CASlider * slider = CASlider::createWithCenter(
        CCRect(size.width * 0.5, size.height * 0.2, size.width * 0.8, 20));
    //绑定监听
    slider->addTarget(this, CAControl_selector(FirstViewController::zoomViewBySliderValue));
    //添加
```

```cpp
        this->getView()->insertSubview(slider, 1);
    //创建CALabel
    CALabel* label = CALabel::createWithCenter(
            CCRect(size.width*0.5, size.height*0.1, size.width*0.8, 40));
    //设置文本
    label->setText(CCString::createWithFormat("ScaleValue:%.02f",slider->getValue())
            ->getCString());
    //设置居中
    label->setTextAlignment(CATextAlignmentCenter);
    label->setVerticalTextAlignmet(CAVerticalTextAlignmentCenter);
    //设置Tag
    label->setTag(3);
    this->getView()->addSubview(label);

}
void FirstViewController::zoomViewBySliderValue(CAControl* control, CCPoint point)
{
    //获得 silder 对象
    CASlider* slider = (CASlider*)control;
    //获得 silder 的当前值
    float zoomValue = slider->getValue();
    //根据 tag 值获得 image
    CAImageView* image = (CAImageView*)this->getView()->getSubviewByTag(1);
    //设置 image 的缩放比
    image->setScale(zoomValue);
    //根据 tag 获得 CALabel
    CALabel* label = (CALabel*)this->getView()->getSubviewByTag(3);
    //设置文本
    label->setText(CCString::createWithFormat("ScaleValue:%.02f",
            slider->getValue())->getCString());
}
```

(2) 代码解析

通过 addTarget 来为 CASlider 绑定一个监听，又在监听函数中，根据 CASlider 的变化来改变 CAImageView 的缩放值，可以应用同样的原理进行其他数值的改变。

3.11 步进控件 CAStepper

CAStepper 是步进控件，它的作用和 CASlider 非常类似，只是 CAStepper 改变固定值，它包括左右两部分，左部为减少，右部为增加，如图 3-11 所示。

CAStepper 的常用函数也与 CASlider 非常类似，如表 3-10 所示。

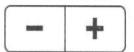

图 3-11 步进控件

表 3-10 CAStepper 常用函数

函 数	说 明
void setValue(float value)	设置当前值
void setMinValue(float minValue)	设置最小值
void setMaxValue(float maxValue)	设置最大值
void setStepValue(double var)	设置步进值
void setAutoRepeat(bool var)	是否支持长按改变值
void setWraps(bool var)	设置是否可在最大值和最小值循环（当增长到最大值时，再单击增大则变成最小值；最小值再减小则变成最大值）
void setBackgroundImage（CAImage * image，CAControlState state）	CAStepper 的背景
void setIncrementImage（CAImage * image，CAControlState state）	右部增加部分的背景
void setDecrementImage（CAImage * image，CAControlState state）	左部减少部分的背景
void setDividerImage（CAImage * image，CAControlState state）	左右部的分割线

下面案例用 CAStepper 去控制一张图片的旋转，每次旋转 30°。

（1）在 FirstViewController.h 中添加一个监听函数，来监听 CAStepper 的值的变化。

```
//监听函数
void stepperValueChange(CAControl* control, CCPoint point);
```

然后在 FirstViewController.cpp 添加 CAStepper 控件并为其绑定监听。
代码如下：

```
void FirstViewController::viewDidLoad()
{
    // Do any additional setup after loading the view from its nib.
    CCRect winRect = this->getView()->getBounds();
    CCSize size = winRect.size;
    //创建 CAImageView 用于选择
    CAImageView* imageView =
        CAImageView::createWithImage(CAImage::create("HelloWorld.png"));
    imageView->setFrame(winRect);
```

```
    imageView->setTag(1);
    this->getView()->addSubview(imageView);
    CAStepper* stepper = CAStepper::createWithCenter(
        CCRect(size.width * 0.5, size.height * 0.6, 200, 200));
    //最大值
    stepper->setMaxValue(360);
    //最小值
    stepper->setMinValue(0);
    //每次变化值(步进值)
    stepper->setStepValue(30);
    //设置监听
    stepper->addTarget(this, CAControl_selector(FirstViewController::stepperValueChange));
    //是否开启触摸特效
    stepper->setTouchEffect(true);
    //是否开始长按效果(flase必须一次一次按,true可以长按改变值。默认为true)
    stepper->setAutoRepeat(true);
    //设置是否可在最大值和最小值循环(当增长到最大值时,再单击增大则变成最小值。最小值再减小则变成最大值)
    stepper->setWraps(true);
    //添加到屏幕
    this->getView()->addSubview(stepper);
}
void FirstViewController::stepperValueChange(CAControl* control, CCPoint point)
{
    //获得stepper对象
    CAStepper* stepper = (CAStepper*)control;
    //根据tag获得imageView
    CAImageView* imageView = (CAImageView*)this->getView()->getSubviewByTag(1);
    //获得stepper的当前值
    float zoomValue = stepper->getValue();
    //设置旋转角度
    imageView->setRotation(zoomValue);
}
```

(2) 代码解析

通过以下代码定义了步进控件的最大值为 360,最小值为 0 以及步进值为 30,分别如下:

```
//最大值
stepper->setMaxValue(360);
//最小值
stepper->setMinValue(0);
//每此变化值(步进值)
stepper->setStepValue(30);
```

然后设置监听如下:

```
stepper -> addTarget(this, CAControl_selector(FirstViewController::stepperValueChange));
```

这样就可以在监听函数中获取 CAStepper 的值来控制 CAImageView 的旋转角度了。

3.12 滚动视图 CAScrollView

CAScrollView 即滚动视图。可支持裁剪其矩形区域以外的渲染,并让其子视图支持滑动操作与缩放功能。在使用 CAScrollView 时,有时需要对 CAScrollView 的一个事件进行监听,那么就需要实现 CAScrollViewDelegate,并重写其函数。

表 3-11 是 CAScrollView 的常用函数。

表 3-11 CAScrollView 常用函数

函　　数	说　　明
void setViewSize(const CCSize& var)	设置容器的内部大小
void setContentOffset(const CCPoint& offset, bool animated)	相对于视图顶部的偏移量
void setBounces(bool var)	设置是否滚动回弹,同时控制水平方向和竖直方向的回弹
void setBounceHorizontal(bool var)	水平方向回弹
void setBounceVertical(bool var)	竖直方向回弹
void setScrollEnabled(bool var)	是否开启滚动
bool isTracking()	是否正在滚动
bool isDecelerating()	是否在惯性运动中
void setShowsHorizontalScrollIndicator(bool var)	设置是否显示水平滚动条
void setShowsVerticalScrollIndicator(bool var)	设置是否显示竖直滚动条
bool isZooming()	是否正在进行缩放控制
void setMinimumZoomScale(float var)	最小缩放比例,默认值为 1
void setMaximumZoomScale(float var)	最大缩放比例,默认值为 1
void setZoomScale(float zoom)	滑动层缩放比例,默认值为 1
void setBackGroundImage(CAImage * image)	设置背景图片
void setBackGroundColor(const CAColor4B &color)	设置背景颜色
void setHeaderRefreshView(CAPullToRefreshView * var)	头部刷新视图
void setFooterRefreshView(CAPullToRefreshView * var)	尾部刷新视图
Void setScrollViewDelegate(CAScrollViewDelegate * var)	滚动事件代理

CAScrollView 相对于前面的控件来说,使用比较复杂。但清楚了每个函数的意义之后,便可以很清晰地构建出我们所需要的 CAScrollView。这里举一个展示文本的 CAScrollView。如图 3-12 所示。

首先,我们希望能够监听到 CAScrollView 的变化,那么就需要去实现 CAScrollViewDelegate。

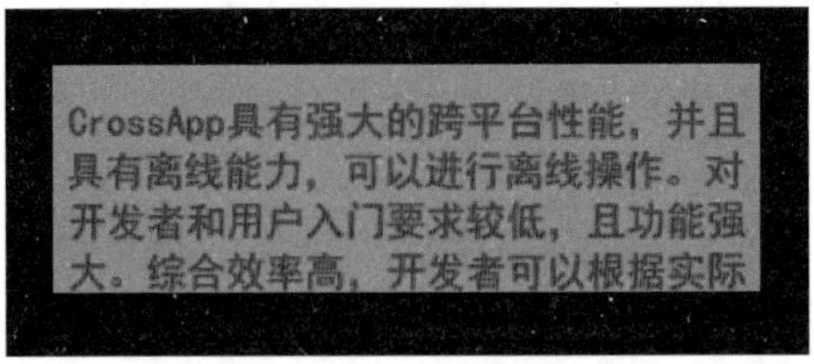

图 3-12　滚动视图

(1) FirstViewController.h 源码如下:

```
#include <iostream>
#include "CrossApp.h"
USING_NS_CC;
class FirstViewController: public CAViewController, public CAScrollViewDelegate
{
public:
    FirstViewController();
    virtual ~FirstViewController();
    //触摸滚动时调用
    virtual void scrollViewDidMoved(CAScrollView* view);
    //触摸滚动停止时调用
    virtual void scrollViewStopMoved(CAScrollView* view);
    //滚动时调用(包括惯性滚动时)
    virtual void scrollViewDidScroll(CAScrollView* view);
    //开始滚动时调用
    virtual void scrollViewWillBeginDragging(CAScrollView* view);
    //结束滚动时调用
    virtual void scrollViewDidEndDragging(CAScrollView* view);
    //缩放时调用
    virtual void scrollViewDidZoom(CAScrollView* view);
    //头部开始刷新时调用
    virtual void scrollViewHeaderBeginRefreshing(CAScrollView* view);
    //尾巴开始刷新时调用
    virtual void scrollViewFooterBeginRefreshing(CAScrollView* view);

protected:
    void viewDidLoad();
    void viewDidUnload();
}
```

(2) 在 FirstViewController.cpp 中去实现这些代理函数:

```
void FirstViewController::viewDidLoad()
{
    // 获得屏幕大小
    CCSize size   = this->getView()->getBounds().size;
```

```cpp
    //设置背景颜色为黑色
    this->getView()->setColor(CAColor_black);
    //创建 ScrollView
    CAScrollView* scrollView = CAScrollView::createWithCenter(
        CCRect(size.width * 0.5, size.height * 0.5 - 270, size.width * 0.5, 100));
    //CAScrollView 容器的大小
    scrollView->setViewSize(CCSize(size.width, 200));
    //设置背景颜色
    scrollView->setBackGroundColor(CAColor_orange);
    //设置背景图片
    //scrollView->setBackGroundImage(CAImage::create("HelloWorld.png"));
    //水平方向是否回弹
    scrollView->setBounceHorizontal(false);
    //竖直方向是否回弹
    scrollView->setBounceVertical(true);
    //是否滚动回弹,控制竖直和水平方向,默认为 ture
    //scrollView->setBounces(false);
    scrollView->setScrollViewDelegate(this);
    this->getView()->addSubview(scrollView);
    //创建 label
    CALabel* label = CALabel::createWithFrame(CCRect(0, 0, size.width * 0.5, 200));
    //设置水平居中
    label->setTextAlignment(CATextAlignmentCenter);
    //设置竖直居中
    label->setVerticalTextAlignmet(CAVerticalTextAlignmentCenter);
    //设置字体大小
    label->setFontSize(18 * CROSSAPP_ADPTATION_RATIO);
    //设置文本内容
    label->setText(UTF8("CrossApp 具有强大的跨平台性能,并且具有离线能力,可以进行离线操作。对于开发者和用户而言,入门要求较低,且功能强大,综合效率高。开发者可以根据实际情况,考虑各方面因素来选择合适的开发解决方案。"));
    //设置文本颜色
    label->setColor(CAColor_blue);
    //将 label 添加到 scrollView
    scrollView->addSubview(label);

}
//触摸滚动时调用
void FirstViewController::scrollViewDidMoved(CAScrollView* view)
{
    CCLog("DidMoved-->");
}
//触摸滚动停止时调用
void FirstViewController::scrollViewStopMoved(CAScrollView* view)
{
```

```
    CCLog("StopMoved-->");
}
//滚动时调用(包括惯性滚动时)
void FirstViewController::scrollViewDidScroll(CAScrollView* view)
{
    CCLog("DidMScroll-->");
}
//开始滚动时调用
void FirstViewController::scrollViewWillBeginDragging(CAScrollView* view)
{
    CCLog("BeginDragging-->");
}
//结束滚动时调用
void FirstViewController::scrollViewDidEndDragging(CAScrollView* view)
{
    CCLog("DidEndDragging-->");
}
//缩放时调用
void FirstViewController::scrollViewDidZoom(CAScrollView* view)
{
    CCLog("DidZoom-->");
}
//头部开始刷新时调用
void FirstViewController::scrollViewHeaderBeginRefreshing(CAScrollView* view)
{
    CCLog("HeaderBeginRefreshing-->");
}
//尾巴开始刷新时调用
void FirstViewController::scrollViewFooterBeginRefreshing(CAScrollView* view)
{
    CCLog("FooterBeginRefreshing-->");
}
```

（3）代码解析

在使用滚动视图过程是将子视图添加到滚动视图，代码如下：

`scrollView->addSubview(label);`

通过如下代码可以设置滚动的回调：

`scrollView->setScrollViewDelegate(this);`

在代理回调的函数里面打印 log，观察控制台打印 log 的时机，也可以在相应的地方实现自己的逻辑。

3.13 列表视图 CAListView

CAListView 和 CAScrollView 非常相似，只是其内部成列表状，支持水平方案和竖直方向的滑动。常用于一些列表信息的展示，如通讯录、新闻列表和目录索引等，如图 3-13 所示。

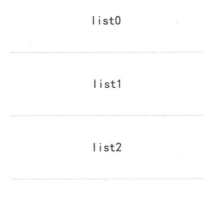

图 3-13 列表控件

CAListView 使用起来相对比较复杂，一般我们要同时使用 CAListView、CAListViewCell、CAListViewDelegate 和 CAListViewDataSource 来构建列表界面，这里先分别了解一下它们的作用。

(1) CAListView 就是列表控件，是显示列表的载体，它是由多个 CAListViewCell 列组成的。

(2) CAListViewCell 是组成列表的每一个单元，下面我们都简称为 cell。

(3) CAListViewDelegate 是 CAListView 的交互代理，主要代理选择 cell 和取消选择 cell 的事件。

(4) CAListViewDataSource 是 CAListView 的数据代理，主要代理 cell 的数量、cell 的高度和将 cell 添加到 CAListView 显示。

表 3-12 为 CAListView 的常用的函数。

表 3-12 CAListView 常用函数

函数	说明
void setListViewOrientation(CAListViewOrientation var)	listView 的滚动方向
void setListHeaderView(CAView * var)	添加头部视图
void setListFooterView(CAView * var)	添加尾部视图
void setListHeaderHeight(unsigned int var)	设置头部视图的高度

续表

函数	说明
void setListFooterHeight(unsigned int var)	设置尾部视图的高度
void setSeparatorColor(CAColor4B var)	设置 cell 分割线的颜色
void setSeparatorViewHeight(unsigned int var)	设置 cell 分割线的高度
void setAllowsSelection(bool var)	是否开启 cell 选择
void setAllowsMultipleSelection(bool var)	是否可以多选 cell
Void setListViewDelegate(CAListViewDelegate * var)	添加交互代理
Void setListViewDataSource(CAListViewDataSource * var)	添加数据代理
void reloadViewSizeData()	刷新 CAListView 的数据
void reloadData()	刷新 CAListView
void setSelectAtIndex(unsigned int index)	根据索引设置 cell 为选中状态
CAListViewCell * dequeueReusableCellWithIdentifier (const char * reuseIdentifier)	从复用队列中寻找指定标识符的 cell

表 3-13 为 CAListViewDelegate 的回调函数。

表 3-13　CAListViewDelegate 回调函数

函数	说明
virtual void listViewDidSelectCellAtIndex(CAListView * listView, unsigned int index)	选中 cell 时调用
virtual void listViewDidDeselectCellAtIndex(CAListView * listView, unsigned int index)	取消选择 cell 时调用

表 3-14 为 CAListViewDataSource 的回调函数。

表 3-14　CAListViewDataSource 回调函数

函数	说明
virtual unsigned int numberOfIndex(CAListView * listView)	cell 的总数量
virtual unsigned int listViewHeightForIndex (CAListView * listView, unsigned int index)	cell 的高度
virtual CAListViewCell * listViewCellAtIndex (CAListView * listView, const CCSize& cellSize, unsigned int index)	添加生成 cell

表 3-15 为 CAListViewCell 的回调函数。

表 3-15　CAListViewCell 回调函数

函数	说明
static CAListViewCell * create (const std::string& reuseIdentifier)	根据重用标示符来创建一个 CAListViewCell
void setBackgroundView(CAView * var)	设置背景

续表

函　　数	说　　明
virtual void normalListViewCell()	正常状态下调用
virtual void highlightedListViewCell()	高亮状态下调用
virtual void selectedListViewCell()	选择状态下调用
virtual void disabledListViewCell()	禁用状态下调用
virtual void recoveryListViewCell()	恢复状态下调用

了解了CAListView的主要函数后，我们来实现一个CAListView的列表视图。

1. 创建我们自己的 cell

我们需要创建一个 cell 的 class，我这里创建一个 MyCell，并继承 CAListViewCell，用于每个列表单元的布局显示。下面给出 MyCell.h 和 MyCell.cpp 的代码实现。

（1）MyCell.h 代码如下：

```cpp
#pragma once
#include "CrossApp.h"
class MyCell : public CAListViewCell
{
public:
    MyCell();
    ~MyCell();
    //创建 MyCell
    static MyCell * create(const std::string& identifier, const CADipRect& _rect = CADipRectZero);
public:
    //初始化 Cell
    void initWithCell();
    //设置回调
    void cellBtnCallback(CAControl * btn, CCPoint point);
protected:
    //正常状态
    virtual void normalListViewCell();
    //高亮状态
    virtual void highlightedListViewCell();
    //选中状态
    virtual void selectedListViewCell();
    //禁用状态
    virtual void disabledListViewCell();
};
```

（2）MyCell.cpp 代码如下：

```cpp
#include "MyCell.h"
MyCell::MyCell()
```

```cpp
{
}
MyCell::~MyCell()
{
}
MyCell* MyCell::create(const std::string& identifier, const CADipRect& _rect)
{
    MyCell* listViewCell = new MyCell();
    //设置重用标示符
    if (listViewCell&&listViewCell->initWithReuseIdentifier(identifier))
    {
        //设置Frame范围
        listViewCell->setFrame(_rect);
        //设置为内存自动释放
        listViewCell->autorelease();
        return listViewCell;
    }
    //如果创建失败则安全释放内存
    CC_SAFE_DELETE(listViewCell);
    return NULL;
}

void MyCell::initWithCell()
{
    //获得当前的宽度
    CADipSize _size = this->getFrame().size;
    //创建CALabel
    CALabel* test = CALabel::createWithCenter(CADipRect(_size.width * 0.5,
        _size.height * 0.5,
        _size.width * 0.8,
        _size.height));
    test->setTextAlignment(CATextAlignmentCenter);
    test->setVerticalTextAlignmet(CAVerticalTextAlignmentCenter);
    test->setFontSize(_px(40));
    test->setTag(100);
    this->addSubview(test);
}
void MyCell::cellBtnCallback(CAControl* btn, CCPoint point)
{
    CCLog("MyCell::cellBtnCallback -->");
}
void MyCell::normalListViewCell()
{
    this->setBackgroundView(CAView::createWithColor(CAColor_white));
}
void MyCell::highlightedListViewCell()
{
```

```cpp
    this->setBackgroundView(CAView::createWithColor(CAColor_yellow));
}
void MyCell::selectedListViewCell()
{
    this->setBackgroundView(CAView::createWithColor(CAColor_orange));
}
void MyCell::disabledListViewCell()
{
    this->setBackgroundView(CAView::createWithColor(CAColor_black));
}
```

2. 实现 CAListViewDelegate 和 CAListViewDataSource 代理

创建了自己的 cell 之后，我们还需要实现 CAListViewDelegate 和 CAListViewDataSource。需要在 FirstViewController 中实现它们，在 FirstViewController.h 中声明下面的函数。

```cpp
#ifndef __HelloCpp__ViewController__
#define __HelloCpp__ViewController__
#include <iostream>
#include "CrossApp.h"
#include "MyCell.h"
USING_NS_CC;
class FirstViewController: public CAViewController, CAListViewDelegate, CAListViewDataSource
{
public:
    FirstViewController();
    virtual ~FirstViewController();
public:
    //选择 cell 时调用
    virtual void listViewDidSelectCellAtIndex(CAListView* listView, unsigned int index);
    //取消选择 cell 时调用
    virtual void listViewDidDeselectCellAtIndex(CAListView* listView, unsigned int index);
    //ListView 中 cell 的数量
    virtual unsigned int numberOfIndex(CAListView* listView);
    //cell 的高度
    virtual unsigned int listViewHeightForIndex(CAListView* listView, unsigned int index);
    //生成 cell
     virtual CAListViewCell* listViewCellAtIndex(CAListView* listView, const CCSize& cellSize, unsigned int index);
protected:
    void viewDidLoad();
    void viewDidUnload();

};
#endif /* defined(__HelloCpp__ViewController__) */
```

3. 实现对应的逻辑

FirstViewController.cpp 源码如下：

```cpp
#include "FirstViewController.h"
FirstViewController::FirstViewController()
{
}
FirstViewController::~FirstViewController()
{
}
void FirstViewController::viewDidLoad()
{
    // Do any additional setup after loading the view from its nib.
    CCRect winRect = this->getView()->getBounds();
    this->getView()->setColor(CAColor_blue);
    //头部视图的 Label
    CALabel* headLabel = CALabel::createWithFrame(
        CCRect(0, 0, winRect.size.width, winRect.size.height / 5));
    headLabel->setText(UTF8("我来组成头部"));
    headLabel->setTextAlignment(CATextAlignmentCenter);
    headLabel->setVerticalTextAlignmet(CAVerticalTextAlignmentCenter);
    //尾部视图的 Label
    CALabel* footLabel = CALabel::createWithFrame(
        CCRect(0, 0, winRect.size.width, winRect.size.height / 5));
    footLabel->setText(UTF8("我来组成腿和脚"));
    footLabel->setTextAlignment(CATextAlignmentCenter);
    footLabel->setVerticalTextAlignmet(CAVerticalTextAlignmentCenter);

    //创建一个全屏的 ListView
    CAListView* listView = CAListView::createWithFrame(winRect);
    //设置交互监听
    listView->setListViewDelegate(this);
    //设置数据监听
    listView->setListViewDataSource(this);
    //是否允许选中
    listView->setAllowsSelection(true);
    //是否允许多选
    listView->setAllowsMultipleSelection(true);
    /* 设置 ListView 的滚动方向
    CAListViewOrientationHorizontal:横向
    CAListViewOrientationVertical:竖向
    */
    listView->setListViewOrientation(CAListViewOrientationVertical);
    //设置头部视图
    listView->setListHeaderView(headLabel);
    //设置尾部视图
    listView->setListFooterView(footLabel);
    //设置头部视图的高度(默认高度为 0,不显示)
    listView->setListHeaderHeight(_px(100));
    //设置尾部视图的高度(默认高度为 0,不显示)
```

```cpp
        listView->setListFooterHeight(_px(100));
        //设置分割线的颜色
        listView->setSeparatorColor(CAColor_yellow);
        //设置分割线的高度
        listView->setSeparatorViewHeight(3);
        this->getView()->addSubview(listView);
}
void FirstViewController::viewDidUnload()
{
        // Release any retained subviews of the main view.
        // e.g. self.myOutlet = nil;
}
//选择 cell 时调用
void FirstViewController::listViewDidSelectCellAtIndex(CAListView * listView, unsigned int index)
{
}
//取消选择 cell 时调用
void FirstViewController::listViewDidDeselectCellAtIndex(CAListView * listView, unsigned int index)
{
}
//ListView 中 cell 的数量
unsigned int FirstViewController::numberOfIndex(CAListView * listView)
{
        return 10;
}
//cell 的高度
unsigned int FirstViewController::listViewHeightForIndex(CAListView * listView, unsigned int index)
{
        return _px(200);
}
//生成 cell
CAListViewCell * FirstViewController:: listViewCellAtIndex (CAListView * listView, const CCSize& cellSize, unsigned int index)
{
        //获得 cell 的高度
        CADipSize _size = cellSize;
        //通过重用标示符获得内存中已有的 MyCell
        MyCell * cell = (MyCell *)listView->dequeueReusableCellWithIdentifier("ListViewCell");
        //如果内存中没有可用的 MyCell,则创建一个新的 cell
        if (!cell)
        {
                cell = MyCell::create("ListViewCell", CADipRect(0, 0, _size.width, _size.height));
                cell->initWithCell();
        }
```

```
        char idx[10] = "";
        sprintf(idx, "list%d", index);
        //获得MyCell的子节点
        CALabel* test = (CALabel*)cell->getSubviewByTag(100);
        //把index显示在Label上
        test->setText(idx);
        return cell;
}
```

这里要特别注意，MyCell是可复用的，即假如列表有100个cell要展示，如果最大只会同时在屏幕中出现5个，那么内存中只会创建5个cell，当下拉滚动需要展示新的cell时则会调用listViewCellAtIndex()函数，从不显示的cell中取出一个，重新复制给cell。所以如果使用的逻辑不够严谨，很容易造成CAListView中的cell顺序展示错位的问题。

3.14 表格视图 CATableView

CATableView主要用于生成列表，区别于CAListView的是增加了分组功能，在table中展示数据，是一个一维的表，可以让用户通过分层的数据进行导航。表可以是静态的或者动态的，可通过dataSource协议和delegate协议实现很多个性化定制，即使拥有大量数据效率也非常高。CATableView只能有一列数据(cell)，且只支持纵向滑动。

首先了解一下CATableView的界面构成。CATableView主要是由两级目录构成，即Selection级和Cell级。如图3-14所示，一个CATextField包含一个或多个Selection，一个Selection包含一个或多个Cell，这样就构成了CATableVIew的层级表示图。

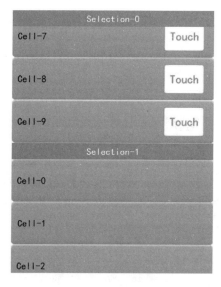

图3-14 表格视图

CATableView 的使用方法和 CAListView 比较类似，也要分别使用 CATableView、CATableViewCell、CATableViewDelegate 和 CATableViewDataSource 来构建。

（1）CATableView 是表格视图的容器，是容器的载体。

（2）CATableViewCell 是表格视图的一个单元（本节后面简称 cell）。

（3）CATableViewDelegate 是交互代理，响应 cell 选中和取消状态。

（4）CATableViewDataSource 是数据代理，设置 Selection 个数及 Selection 包含的 cell 个数。

表 3-16 为 CATableView 的常用函数。

表 3-16　CATableView 常用函数

函　　数	说　　明
void setTableHeaderView(CAView * var)	添加头部视图
void setTableFooterView(CAView * var)	添加尾部视图
void setTableHeaderHeight(unsigned int var)	设置头部视图的高度
void setTableFooterHeight(unsigned int var)	设置尾部视图的高度
void setSeparatorColor(CAColor4B var)	设置 cell 分割线的颜色
void setSeparatorViewHeight(unsigned int var)	设置 cell 分割线的高度
void setAllowsSelection(bool var)	是否开启 cell 选择
void setAllowsMultipleSelection(bool var)	是否可以多选 cell
Void setTableViewDelegate(CATableViewDelegate * var)	添加交互代理
Void setTableViewDataSource(CATableViewDataSource * var)	添加数据代理
void reloadViewSizeData()	刷新 CATableView
void reloadData()	刷新 CATableView 的数据
void setSelectAtIndex(unsigned int index)	根据索引设置 cell 为选中状态
CATableViewCell * dequeueReusableCellWithIdentifier(const char * reuseIdentifier)	从复用队列中寻找指定标识符的 cell

表 3-17 为 CATableViewCell 的常用函数。

表 3-17　CATableViewCell 常用函数

函　　数	说　　明
static CATableViewCell * create(const std::string& reuseIdentifier)	根据重用标示符来创建一个 CAListViewCell
void setBackGroundViewForState(CAControlState controlState, CAView * var)	设置背景
virtual void normalTableViewCell()	正常状态下调用
virtual void highlightedTableViewCell()	高亮状态下调用
virtual void selectedListViewCell()	选择状态下调用
virtual void disabledTableViewCell()	禁用状态下调用
virtual void recoveryTableViewCell()	恢复状态下调用

表 3-18 为 CATableViewDelegate 的常用函数。

表 3-18 CATableViewDelegate 常用函数

函 数	说 明
virtual void tableViewDidSelectRowAtIndexPath(CATableView * table, unsigned int section, unsigned int row)	选中 cell 时调用
virtual void tableViewDidDeselectRowAtIndexPath(CATableView * table, unsigned int section, unsigned int row)	取消选择 cell 时调用

表 3-19 为 CATableViewDataSource 的常用函数。

表 3-19 CATableViewDataSource 常用函数

函 数	说 明
virtual CATableViewCell * tableCellAtIndex(CATableView * table, const CCSize& cellSize, unsigned int section, unsigned int row)	获取指定的 cell
virtual unsigned int numberOfRowsInSection(CATableView * table, unsigned int section)	获取对应的 section 所包含的 cell 个数
virtual unsigned int numberOfSections(CATableView * table)	获取 tableview 包含的 section 个数
virtual unsigned int tableViewHeightForRowAtIndexPath(CATableView * table, unsigned int section, unsigned int row)	获取指定的 cell 高度
virtual unsigned int tableViewHeightForHeaderInSection(CATableView * table, unsigned int section)	获取指定的 section 的 header view 的高度
virtual unsigned int tableViewHeightForFooterInSection(CATableView * table, unsigned int section)	获取指定的 section 的 footer view 的高度

下面是使用 CATableView 来实现一个简单的表单视图，首先需要新建一个 Class，命名为 MyTableViewCell 并继承 CATableViewCell，代码示例如下。

（1）CATableViewCell.h 代码如下：

```
#ifndef _My_TableViewCell_h_
#define _My_TableViewCell_h_
#include <iostream>
#include "CrossApp.h"
USING_NS_CC;
class MyTableViewCell:public CATableViewCell
{
public:
    MyTableViewCell();
    virtual ~MyTableViewCell();
    //创建 MyTableViewCell
    static MyTableViewCell * create(const std::string& identifier, const CADipRect& _rect =
```

```cpp
    CADipRectZero);
public:
    //初始化
    void initWithCell();
    //按钮的回调函数
    void cellBtnCallback(CAControl* btn, CCPoint point);

protected:
    //正常状态下调用
    virtual void normalTableViewCell();
    //高亮状态下调用
    virtual void highlightedTableViewCell();
    //选择状态下调用
    virtual void selectedTableViewCell();
    //禁用状态下调用
    virtual void disabledTableViewCell();
    //恢复状态下调用
    virtual void recoveryTableViewCell();
};
#endif
```

代码解析：

以上的类实现了表格每个条目的定义。

(2) CATableViewCell.cpp 代码如下：

```cpp
#include "MyTableViewCell.h"
MyTableViewCell::MyTableViewCell()
{
}
MyTableViewCell::~MyTableViewCell()
{
}
MyTableViewCell* MyTableViewCell::create(const std::string& identifier, const CADipRect& _rect)
{
    //创建
    MyTableViewCell* tableViewCell = new MyTableViewCell();
    if(tableViewCell&&tableViewCell->initWithReuseIdentifier(identifier))
    {
        tableViewCell->setFrame(_rect);
        tableViewCell->autorelease();
        return tableViewCell;
    }
    CC_SAFE_DELETE(tableViewCell);
    return NULL;
}
```

```cpp
void MyTableViewCell::initWithCell()
{
    //Cell 的大小
    CADipSize m_size = this->getFrame().size;
    //创建 CALabel
    CALabel * cellText = CALabel::createWithCenter(
        CADipRect(m_size.width * 0.1, m_size.height * 0.5, m_size.width * 0.3, m_size.height * 0.8));
    //设置 tag
    cellText->setTag(100);
    //设置字体大小
    cellText->setFontSize(_px(30));
    //设置中心对齐
    cellText->setTextAlignment(CATextAlignmentCenter);
    cellText->setVerticalTextAlignmet(CAVerticalTextAlignmentCenter);
    //添加到当前 cell
    this->addSubview(cellText);
    //创建 CAButton
    CABton * btnOnCell = CAButton::createWithCenter(CADipRect(m_size.width * 0.85,
        m_size.height * 0.5, m_size.width * 0.2, m_size.height * 0.7), CAButtonTypeRoundedRect);
    //设置 tag
    btnOnCell->setTag(102);
    //设置显示文本
    btnOnCell->setTitleForState(CAControlStateAll, "Touch");
    //添加回调监听
    btnOnCell->addTarget(this, CAControl_selector(MyTableViewCell::cellBtnCallback),
        CAControlEventTouchUpInSide);
    //添加到 cell
    this->addSubview(btnOnCell);
}
void MyTableViewCell::cellBtnCallback(CAControl * btn, CCPoint point)
{
    //按钮被单击时打印 log
    CCLog("MyTableViewCell::cellBtnCallback -->");
}
void MyTableViewCell::normalTableViewCell()
{
    //改变背景颜色
    this->setBackgroundView(CAView::createWithColor(CAColor_white));
}
void MyTableViewCell::highlightedTableViewCell()
{
    //改变背景颜色
    this->setBackgroundView(CAView::createWithColor(CAColor_gray));
}
```

```cpp
void MyTableViewCell::selectedTableViewCell()
{
    //改变背景颜色
    this->setBackgroundView(CAView::createWithColor(CAColor_orange));
}
void MyTableViewCell::disabledTableViewCell()
{
    //改变背景颜色
    this->setBackgroundView(CAView::createWithColor(CAColor_black));
}
void MyTableViewCell::recoveryTableViewCell()
{
    //改变背景颜色
    this->setBackgroundView(CAView::createWithColor(CAColor_blue));
}
```

代码解析：

以上代码创建了每个表格的实现定义，定义了表格的初始化和在不同状态下对应的背景变化。

```cpp
void MyTableViewCell::normalTableViewCell()         //默认状态调用该函数
void MyTableViewCell::highlightedTableViewCell()    //选中之后
void MyTableViewCell::selectedTableViewCell()       //选择时候
void MyTableViewCell::disabledTableViewCell()       //无效时候
void MyTableViewCell::recoveryTableViewCell()       //恢复时候
```

下面在 FirstViewController 中实现 CATableViewDelegate 和 CATableViewDataSource 这两个代理。

（1）FirstViewController.h 的代码如下：

```cpp
#ifndef __HelloCpp__ViewController__
#define __HelloCpp__ViewController__
#include <iostream>
#include "CrossApp.h"
#include "MyTableViewCell.h"
USING_NS_CC;
class FirstViewController: public CAViewController, public CATableViewDataSource,
public CATableViewDelegate
{
public:
    FirstViewController();
    virtual ~FirstViewController();
public:
    //选中 cell 时触发
    virtual void tableViewDidSelectRowAtIndexPath(CATableView* table,
                  unsigned int section, unsigned int row);
    //取消选中 cell 时触发
```

```cpp
        virtual void tableViewDidDeselectRowAtIndexPath(CATableView* table,
                    unsigned int section, unsigned int row);
    //获取对应的section包含的cell个数
    virtual unsigned int numberOfRowsInSection(CATableView * table, unsigned int section);
    //获取tableview包含的section个数
    virtual unsigned int numberOfSections(CATableView * table);
    //获得指定cell
    virtual CATableViewCell* tableCellAtIndex(CATableView* table,
            const CCSize& cellSize, unsigned int section, unsigned int row);
    //设置section的头部
            virtual CAView* tableViewSectionViewForHeaderInSection(CATableView* table,
            const CCSize& viewSize, unsigned int section);
    //设置section的尾部
    virtual CAView* tableViewSectionViewForFooterInSection(CATableView* table,
            const CCSize& viewSize, unsigned int section);
    //获取指定的cell高度
    virtual unsigned int tableViewHeightForRowAtIndexPath(CATableView* table,
            unsigned int section, unsigned int row);
    //获得指定的section的header view的高度
    virtual unsigned int tableViewHeightForHeaderInSection(CATableView* table,
            unsigned int section);
    //获得指定的section的footer view的高度
    virtual unsigned int tableViewHeightForFooterInSection(CATableView* table,
            unsigned int section);
protected:
    void viewDidLoad();
    void viewDidUnload();
};
#endif /* defined(__HelloCpp__ViewController__) */
```

（2）FirstViewController.cpp中实现表格视图如下：

```cpp
#include "FirstViewController.h"
FirstViewController::FirstViewController()
{
}
FirstViewController::~FirstViewController()
{
}
void FirstViewController::viewDidLoad()
{
    CCSize size = this->getView()->getBounds().size;
    CATableView* p_TableView = CATableView::createWithCenter(CADipRect(size.width * 0.5,
            size.height * 0.5, size.width, size.height));
    p_TableView->setTableViewDataSource(this);
    p_TableView->setTableViewDelegate(this);
    p_TableView->setAllowsSelection(true);              //设置可以选择
```

```cpp
    p_TableView->setAllowsMultipleSelection(true);   //设置可以选择多项
    p_TableView->setSeparatorColor(CAColor_clear);
    this->getView()->addSubview(p_TableView);
}

void FirstViewController::viewDidUnload()
{
    // Release any retained subviews of the main view.
    // e.g. self.myOutlet = nil;
}

void FirstViewController::tableViewDidSelectRowAtIndexPath(CATableView* table,
      unsigned int section, unsigned int row)
{
}
void FirstViewController::tableViewDidDeselectRowAtIndexPath(CATableView* table, unsigned int section, unsigned int row)
{
}

CATableViewCell* FirstViewController::tableCellAtIndex(CATableView* table,
    const CCSize& cellSize, unsigned int section, unsigned int row)
{
    CADipSize _size = cellSize;
    //根据标识获得 cell
    MyTableViewCell* cell =
    dynamic_cast<MyTableViewCell*>(
           table->dequeueReusableCellWithIdentifier("CrossApp"));
    //如果没有找到 cell
    if (cell == NULL)
    {
        //创建一个
        cell = MyTableViewCell::create("CrossApp", CADipRect(0, 0, _size.width, _size.height));
        //调用 cell 的初始化
        cell->initWithCell();
    }
    //如果是 section 为 1 的情况
    if (section == 1)
    {
        //根据 tag 获得 CAButton
        CAButton* cellBtn = (CAButton*)cell->getSubviewByTag(102);
        //隐藏按钮
        cellBtn->setVisible(false);
    }
    else
```

```cpp
    {
        //根据 tag 获得 CAButton
        CAButton * cellBtn = (CAButton * )cell->getSubviewByTag(102);
        //显示按钮
        cellBtn->setVisible(true);
    }
    CCString * order = CCString::createWithFormat("Cell-%d",row);
    //根据 tag 获得 CALabel
    CALabel * cellText = (CALabel * )cell->getSubviewByTag(100);
    //设置文本显示
    cellText->setText(order->getCString());
    return cell;
}
CAView * FirstViewController::tableViewSectionViewForHeaderInSection(CATableView * table,
    const CCSize& viewSize, unsigned int section)
{
    CCLog("Header -->");
    CCString * head = CCString::createWithFormat("Selection-%d", section);
    //创建 Section 头部视图
    CAView * view = CAView::createWithColor(CAColor_gray);
    CADipSize _size = viewSize;
    CALabel * header = CALabel::createWithCenter(CADipRect(_size.width * 0.5, _size.height * 0.5,
        _size.width * 0.8, _size.height));
    header->setText(head->getCString());
    header->setFontSize(_px(30));
    header->setColor(CAColor_white);
    header->setTextAlignment(CATextAlignmentCenter);
    header->setVerticalTextAlignmet(CAVerticalTextAlignmentCenter);
    view->addSubview(header);
    return view;
}
CAView * FirstViewController::tableViewSectionViewForFooterInSection(CATableView * table,
    const CCSize& viewSize, unsigned int section)
{
    CCLog("Footer -->");
    CAView * view = CAView::createWithColor(CAColor_white);
    return view;
}

unsigned int FirstViewController::numberOfRowsInSection(CATableView * table,
    unsigned int section)
{
    //cell 数,从 0 计算。10 表示 0~9
    return 10;
```

```cpp
}
unsigned int FirstViewController::numberOfSections(CATableView * table)
{
    //表格数,从 0 开始计算。4 则表示 0~3
    return 4;
}
unsigned int FirstViewController::tableViewHeightForRowAtIndexPath(CATableView * table,
    unsigned int section, unsigned int row)
{
    //高度
    return _px(130);
}

unsigned int FirstViewController::tableViewHeightForHeaderInSection(CATableView * table,
    unsigned int section)
{
    //高度
    return _px(50);
}
unsigned int FirstViewController::tableViewHeightForFooterInSection(CATableView * table,
    unsigned int section)
{
    return 1;
}
```

代码解析:

在 FirstViewController 类中通过回调函数定义了表格的分组以及每个分组的表格条目数量,一般在开发实际项目的时候,会结合集合对象来完成分组数据和条目数据的返回,在 tableCellAtIndex 函数中定义了每个条目的视图对象,以下代码是为了复用每个列表条目:

```cpp
//根据标识获得 cell
MyTableViewCell * cell =
dynamic_cast < MyTableViewCell * >(
        table->dequeueReusableCellWithIdentifier("CrossApp"));
//如果没有找到 cell
if (cell == NULL)
{   cell = MyTableViewCell::create("CrossApp", CADipRect(0, 0, _size.width, _size.height));
    cell->initWithCell();
}
```

这样就构建了一个表格视图,读者可以自己尝试用 CATableView 去实现微信的通讯录列表。

3.15 容器 CACollectionView

CACollectionView 与 CATableView 类似，主要用于数据的展示，实现了 tableView 的基本功能，同时对 tableView 拓展，更完美地进行数据展示。容器视图如图 3-15 所示。

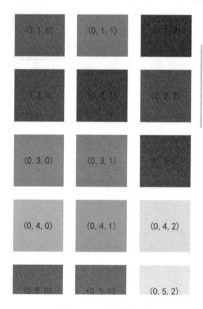

图 3-15　容器视图

CACollectionView 的使用方法和 CATableView 比较类似，也要分别使用 CACollectionView、CACollectionViewCell、CACollectionViewDelegate 和 CACollectionViewDataSource 来构建。

(1) CACollectionView 是表格视图的容器，是容器的载体。

(2) CACollectionViewCell 是表格视图的一个单元(本节后面简称 cell)。

(3) CACollectionViewDelegate 是交互代理，响应 cell 选中和取消状态。

(4) CACollectionViewDataSource 是数据代理，设置 Selection 个数及 Selection 包含 cell 个数。

表 3-20 为 CACollectionView 的常用函数。

表 3-20　CACollectionView 的常用函数

函　　数	说　　明
void setCollectionHeaderView(CAView * var)	添加头部视图
void setCollectionFooterView(CAView * var)	添加尾部视图
void setCollectionHeaderHeight(unsigned int var)	设置头部视图的高度
void setCollectionFooterHeight(unsigned int var)	设置尾部视图的高度
unsigned int setHoriInterval(unsigned int var)	item 间的水平间隔

续表

函数	说明
unsigned int setVertInterval(unsigned int var)	item 间的竖直间隔
void setAllowsSelection(bool var)	是否开启 cell 选择
void setAllowsMultipleSelection(bool var)	是否可以多选 cell
Void setTableViewDelegate(CATableViewDelegate * var)	添加交互代理
Void setTableViewDataSource(CATableViewDataSource * var)	添加数据代理
void reloadData()	刷新 CATableView 的数据
void setSelectAtIndex(unsigned int index)	根据索引设置 cell 为选中状态
CATableViewCell * dequeueReusableCellWithIdentifier (const char * reuseIdentifier)	从复用队列中寻找指定标识符的 cell

表 3-21 为 CACollectionViewDelegate 的常用函数。

表 3-21　CACollectionViewDelegate 常用函数

函数	说明
virtual void collectionViewDidSelectRowAtIndexPath（CACollectionView * collectionView，unsigned int section，unsigned int item）	选中 cell 时调用
virtual void collectionViewDidDeselectRowAtIndexPath（CACollectionView * collectionView，unsigned int section，unsigned int item）	取消选择 cell 时调用

表 3-22 为 CACollectionDataSource 的常用函数。

表 3-22　CACollectionDataSource 常用函数

函数	说明
virtual CACollectionViewCell * collectionCellAtIndex（CACollectionView * collectionView，const CCSize& cellSize，unsigned int section，unsigned int row，unsigned int item）	获取指定 cell
virtual unsigned int numberOfRowsInSection(CACollectionView * collectionView，unsigned int section)	获取对应的 section 所包含的 cell 个数
virtual unsigned int numberOfSections (CACollectionView * collectionView)	获取 tableview 包含的 section 个数
virtual unsigned int collectionViewHeightForHeaderInSection (CACollectionView * collectionView，unsigned int section)	每个 section 的 headerView
virtual unsigned int collectionViewHeightForFooterInSection (CACollectionView * collectionView，unsigned int section)	每个 section 的 footerView
virtual CAView * collectionViewSectionViewForHeaderInSection(CACollectionView * collectionView，const CCSize& viewSize，unsigned int section)	headerView 的内容
virtual CAView * collectionViewSectionViewForFooterInSection(CACollectionView * collectionView，const CCSize& viewSize，unsigned int section)	footerView 的内容

续表

函　数	说　明
virtual unsigned int numberOfItemsInRowsInSection（CACollectionView ＊ collectionView，unsigned int section，unsigned int row）	每个 cell 里的 item 数量
virtual unsigned int collectionViewHeightForRowAtIndexPath（CACollectionView ＊ collectionView，unsigned int section，unsigned int row）	cell 的高度

下面是本节的示例代码。

（1）FirstViewController.h 内容：

```cpp
#ifndef __HelloCpp__ViewController__
#define __HelloCpp__ViewController__
#include <iostream>
#include "CrossApp.h"
USING_NS_CC;
class FirstViewController : public CAViewController, CACollectionViewDelegate, CACollectionViewDataSource
{
public:
    FirstViewController();
    virtual ~FirstViewController();
protected:
    void viewDidLoad();
    void viewDidUnload();
public:
    //选中 item 时调用
    virtual void collectionViewDidSelectCellAtIndexPath(CACollectionView * collectionView,
        unsigned int section, unsigned int row, unsigned int item);
    //取消 item 时调用
    virtual void collectionViewDidDeselectCellAtIndexPath(CACollectionView * collectionView,
        unsigned int section, unsigned int row, unsigned int item);
    //获取指定 cell
    virtual CACollectionViewCell * collectionCellAtIndex(CACollectionView * collectionView,
     const CCSize& cellSize, unsigned int section, unsigned int row, unsigned int item);
    //section 的个数
    virtual unsigned int numberOfSections(CACollectionView * collectionView);
    //section 中的 cell 个数
    virtual unsigned int numberOfRowsInSection(CACollectionView * collectionView,
     unsigned int section);
    //每个 cell 中 item 的个数
    virtual unsigned int numberOfItemsInRowsInSection(CACollectionView * collectionView,
        unsigned int section, unsigned int row);
    //cell 的高度
    virtual unsigned int collectionViewHeightForRowAtIndexPath(
    CACollectionView * collectionView, unsigned int section, unsigned int row);
```

```
private:
    //用于获得屏幕的size
    CADipSize size;
    //CACollectionView
    CACollectionView* p_Conllection;
    //颜色容器
    std::vector<CAColor4B> colorArr;
};
#endif /* defined(__HelloCpp__ViewController__) */
```

(2) FirstViewController.cpp 内容：

```
#include "FirstViewController.h"
FirstViewController::FirstViewController()
{
}
FirstViewController::~FirstViewController()
{
}
void FirstViewController::viewDidLoad()
{
    //获得屏幕大小
    size = this->getView()->getBounds().size;
    //随机出颜色
    for (int i = 0; i < 40; i++)
    {
        char r = CCRANDOM_0_1() * 255;
        char g = CCRANDOM_0_1() * 255;
        char b = CCRANDOM_0_1() * 255;
        //将随机的ccc4对象放入到容器里
        colorArr.push_back(ccc4(r, g, b, 255));
    }
    //生成CACollectionView
    p_Conllection = CACollectionView::createWithFrame(this->getView()->getBounds());
    //开启选中
    p_Conllection->setAllowsSelection(true);
    //开启多选
    p_Conllection->setAllowsMultipleSelection(true);
    //绑定交互代理
    p_Conllection->setCollectionViewDelegate(this);
    //绑定数据代理
    p_Conllection->setCollectionViewDataSource(this);
    //item水平间的距离
    p_Conllection->setHoriInterval(_px(40));
    //item竖直间的距离
    p_Conllection->setVertInterval(_px(40));
    //添加到屏幕渲染
```

```cpp
    this->getView()->addSubview(p_Conllection);
}
void FirstViewController::viewDidUnload()
{
    // Release any retained subviews of the main view.
    // e.g. self.myOutlet = nil;
}
void FirstViewController::collectionViewDidSelectCellAtIndexPath(CACollectionView * collectionView,
unsigned int section, unsigned int row, unsigned int item)
{
    //选中
    CCLog(UTF8("选中"));
}

void FirstViewController::collectionViewDidDeselectCellAtIndexPath(CACollectionView * collectionView,
unsigned int section, unsigned int row, unsigned int item)
{
    //取消选中
    CCLog(UTF8("取消选中"));
}

CACollectionViewCell * FirstViewController::collectionCellAtIndex(
    CACollectionView * collectionView, const CCSize& cellSize,
        unsigned int section, unsigned int row, unsigned int item)
{
    //计算(如果cell个数大于颜色数组,则返回空)
    if (row * 3 + item >= colorArr.size())
    {
        return NULL;
    }
    //获得
    CADipSize _size = cellSize;
    //根据标识获得CACollectionViewCell
    CACollectionViewCell * p_Cell =
      collectionView->dequeueReusableCellWithIdentifier("CrossApp");
    //如果没有找到相应的CACollectionViewCell则新建一个
    if (p_Cell == NULL)
    {
        p_Cell = CACollectionViewCell::create("CrossApp");
        //生成item背景
        CAView * itemImage = CAView::createWithFrame(CADipRect(0, 0, _size.width,
                            _size.height));
        itemImage->setTag(99);
        p_Cell->addSubview(itemImage);
        CADipSize itemSize = itemImage->getBounds().size;
        //生成itemCALabel
        CALabel * itemText = CALabel::createWithCenter(CADipRect(itemSize.width * 0.5,
```

```
                            itemSize.height * 0.5, 150, 40));
            itemText->setTag(100);
            itemText->setFontSize(_px(29));
            itemText->setTextAlignment(CATextAlignmentCenter);
            itemText->setVerticalTextAlignmet(CAVerticalTextAlignmentCenter);
            itemImage->addSubview(itemText);
        }
        //设置item背景颜色
        CAView* itemImageView = p_Cell->getSubviewByTag(99);
        itemImageView->setColor(colorArr.at(row * 3 + item));
        CCLog("%d", row * 3 + item);
        //设置item文本显示
        char pos[20] = "";
        sprintf(pos, "(%d,%d,%d)", section, row, item);
        CALabel* itemText = (CALabel*)p_Cell->getSubviewByTag(99)->getSubviewByTag(100);
        itemText->setText(pos);

        return p_Cell;
}
unsigned int FirstViewController::numberOfSections(CACollectionView* collectionView)
{
        return 1;
}

unsigned int FirstViewController::numberOfRowsInSection(CACollectionView* collectionView,
    unsigned int section)
{
        return colorArr.size() % 3 == 0 ? colorArr.size() / 3 : colorArr.size() / 3 + 1;
}

unsigned int FirstViewController::numberOfItemsInRowsInSection(CACollectionView* collectionView,
unsigned int section, unsigned int row)
{
        return 3;
}

unsigned int FirstViewController::collectionViewHeightForRowAtIndexPath(CACollectionView*
collectionView, unsigned int section, unsigned int row)
{
        return (this->getView()->getBounds().size.width - _px(40) * 4) / 3;
}
```

代码解析:

该案例并没有自定义每个条目,而是使用了CACollectionViewCell定义每个条目,在使用CACollectionView过程中,最主要的是在每一行中增加了多列的处理,可以通过下面函数来为每个条目规划内容:

```
collectionCellAtIndex(
CACollectionView * collectionView, const CCSize& cellSize,
    unsigned int section, unsigned int row, unsigned int item)
```

其中 row 定义了行，item 定义了列，这样就创建了一个多彩的 CACollectionView。

3.16　切换页面 CAPageView

CAPageView 用于实现分页和翻页效果。CAPageView 提供了横向和竖直两个方向的样式，也可以通过继承 CAPageViewDelegate 来实现对 CAPageView 的监听。

表 3-23 为 CAPageView 的常用函数。

表 3-23　CAPageView 的常用函数

函　数　名	说　　明
void setViews(const CAVector<CAView *>& vec)	添加存放 View 的 CAVector 容器
int getPageCount()	CAPgaeView 的总页数

下面介绍代码实例。

（1）首先我们需要继承 CAPageViewDelegate 实现监听函数。在 FirstViewController.h 文件中添加如下代码：

```
#include <iostream>
#include "CrossApp.h"
USING_NS_CC;
class FirstViewController : public CAViewController, public  CAPageViewDelegate
{
public:
    FirstViewController();
    virtual ~FirstViewController();
    //切换开始时调用
    virtual void pageViewDidBeginTurning(CAPageView * pageView);
    //切换结束时调用
    virtual void pageViewDidEndTurning(CAPageView * pageView);
    //选择当前的切换页时调用
     virtual void pageViewDidSelectPageAtIndex(CAPageView * pageView, unsigned int index, const CCPoint& point);
protected:
    void viewDidLoad();
    void viewDidUnload();
};
```

（2）在 FirstViewController.cpp 中实现逻辑：

```
void FirstViewController::viewDidLoad()
```

```cpp
{
    //声明一个CAVector作为添加到PageView的容器
    CAVector<CAView*> viewVector;
    //获得屏幕的rect
    CCRect winRect = this->getView()->getBounds();
    //创建CALabel
    CALabel* labelView = CALabel::createWithFrame(winRect);
    //设置为居中
    labelView->setTextAlignment(CATextAlignmentCenter);
    labelView->setVerticalTextAlignmet(CAVerticalTextAlignmentCenter);
    labelView->setFontSize(75);
    //设置文本内容
    labelView->setText(UTF8("第一个View"));
    //创建CAImageView
    CAImageView* imageView =
        CAImageView::createWithImage(CAImage::create("HelloWorld.png"));
    //设置显示区域
    imageView->setFrame(winRect);
    //创建一个蓝色的View
    CAView* view = CAView::createWithColor(CAColor_blue);
    //设置显示区域
    view->setFrame(winRect);
    //创建CALabel
    CALabel* lastLabel = CALabel::createWithFrame(winRect);
    //设置居中
    lastLabel->setTextAlignment(CATextAlignmentCenter);
    lastLabel->setVerticalTextAlignmet(CAVerticalTextAlignmentCenter);
    lastLabel->setFontSize(75);
    //设置显示文本
    lastLabel->setText(UTF8("最后一个View"));
    //将lastLabel添加到View
    view->addSubview(lastLabel);
    //将上面的三个控件放入到CAVector容器内
    viewVector.pushBack(labelView);
    viewVector.pushBack(imageView);
    viewVector.pushBack(view);
    /*创建一个CAPageView并设置为水平滚动
        CAPageViewDirectionHorizontal:水平
        CAPageViewDirectionVertical:竖直
    */
    CAPageView* pageViewTest =
        CAPageView::createWithCenter(CADipRect(winRect.size.width * 0.5,
        winRect.size.height * 0.5, winRect.size.width, winRect.size.height),
        CAPageView::CAPageViewDirectionHorizontal);
    //设置监听
    pageViewTest->setPageViewDelegate(this);
    //将CAVector添加到pageViewTest
```

```
    pageViewTest->setViews(viewVector);
    pageViewTest->getPageCount();
    //将 pageViewTest 添加到屏幕显示
    this->getView()->addSubview(pageViewTest);
}
void FirstViewController::viewDidUnload()
{
    // Release any retained subviews of the main view.
    // e.g. self.myOutlet = nil;
}
void FirstViewController::pageViewDidBeginTurning(CAPageView * pageView)
{
    CCLog("Begin --->");
}
void FirstViewController::pageViewDidEndTurning(CAPageView * pageView)
{
    CCLog("End --->");
}
void FirstViewController::pageViewDidSelectPageAtIndex(CAPageView * pageView, unsigned int index, const CCPoint& point)
{
    CCLog("Index:%d",index);
}
```

代码解析:

以上代码创建了一个含有 3 个 CAView 的 CAPageView,横向滑动可以切换到不同的 CAView,在切换时会调用 pageViewDidBeginTurning 函数和 pageViewDidEndTurning 函数,当单击某个页面的时候 pageViewDidSelectPageAtIndex 会被调用。

第 4 章 CrossApp 数据存储与解析

在开发 App 时，需要对数据持久保持和解析，CrossApp 也对常见的数据类型解析提供了支持如 JSON、XML 和 SQLite3 等。

4.1 CAUserDefault 简单存储

CrossApp 中提供了自带的存储类即 CAUserDefault，适合存储数据量比较小，结构比较简单的数据，如果需要存储大量的复杂数据，建议使用 SQLite。

表 4-1 为 CAUserDefault 的常用函数。

表 4-1 CAUserDefault 的常用函数

方 法 名	说 明
void setBoolForKey(const char * pkey,bool valuer)	根据 pkey 存储一个 bool 类型
void setIntegerForKey(const char * pkey,int valuer)	根据 pkey 存储一个 int 类型
void setFloatForKey(const char * pkey,float valuer)	根据 pkey 存储一个 float 类型
void setDoubleForKey(const char * pkey,double valuer)	根据 pkey 存储一个 double 类型
void setStringForKey(const char * pkey, const std::string & value)	根据 pkey 存储一个 string 类型
bool getBoolForKey(const char * pKey)	根据 pkey 读取相应的值，如果没有在 UserDefault.xml 找出则返回 false
bool getBoolForKey(const char * pKey,bool defaultValue)	根据 pkey 读取相应的值，如果没有在 UserDefault.xml 找出则返回 defaultValue
int getIntegerForKey(const char * pKey)	根据 pkey 读取相应的值，如果没有在 UserDefault.xml 找出则返回 0
int getIntegerForKey(const char * pKey, int defaultValue)	根据 pkey 读取相应的值，如果没有在 UserDefault.xml 找出则返回 defaultValue
float getFloatForKey(const char * pKey)	根据 pkey 读取相应的值，如果没有在 UserDefault.xml 找出则返回 0.0f
float getFloatForKey(constchar * pKey,float defaultValue)	根据 pkey 读取相应的值，如果没有在 UserDefault.xml 找出则返回 defaultValue

续表

方 法 名	说 明
double getDoubleForKey(const char * pKey)	根据 pkey 读取相应的值,如果没有在 UserDefault.xml 找出则返回 0.0
double getDoubleForKey(const cha * pKey, double defaultValue)	根据 pkey 读取相应的值,如果没有在 UserDefault.xml 找出则返回 defaultValue
void flush()	存储到 UserDefault.xml,不写则不会存入
const string& getXMLFilePath()	获得 UserDefault.xml 的存储路径

(1) 存储数据代码如下:

```
//存储名字
CAUserDefault::sharedUserDefault()->setStringForKey("name", "zero");
//存储年龄
CAUserDefault::sharedUserDefault()->setIntegerForKey("age", 25);
//存储身高
CAUserDefault::sharedUserDefault()->setDoubleForKey("height",1.75);
//存储体重
CAUserDefault::sharedUserDefault()->setFloatForKey("wight", 75.0f);
//存储性别,true 为男,fasle 女
CAUserDefault::sharedUserDefault()->setBoolForKey("sex", true);
//这里一定要提交写入,否则不会记录到 XML 文件中,下次启动游戏时就获取不到 value
CAUserDefault::sharedUserDefault()->flush();
```

将看到 UserDefault.xml 文件如图 4-1 所示。

```
<?xml version="1.0" encoding="UTF-8"?>
<userDefaultRoot>
    <name>zero</name>
    <age>25</age>
    <height>1.750000</height>
    <weight>75.000000</weight>
    <sex>true</sex>
</userDefaultRoot>
```

图 4-1　使用 CAUserDefault 保存的文件

(2) 读取示例代码:

```
//读取 name
std::string name = CAUserDefault::sharedUserDefault()->getStringForKey("name");
//读取 Email,如果没有 Email 则返回"default"
std::string email = CAUserDefault::sharedUserDefault()->getStringForKey("Email", "default");
//读取 age
int age = CAUserDefault::sharedUserDefault()->getIntegerForKey("age");
//读取 id,如果没有 id 则返回 10000
int id = CAUserDefault::sharedUserDefault()->getIntegerForKey("id", 10000);
//读取身高体重,如果没有找到则返回 0
```

```
float height = CAUserDefault::sharedUserDefault()->getFloatForKey("height");
double weight = CAUserDefault::sharedUserDefault()->getDoubleForKey("weight");
//读取 sex,如果没有找到则返回 true
bool sex = CAUserDefault::sharedUserDefault()->getBoolForKey("false", true);
```

4.2 SQLite 的使用

在 CrossApp 中,简单数据的存储可以使用 CAUserDefault。存储大量且不规则的数据可以使用 SQLite 数据库。SQLite 是使用非常广泛的嵌入式数据库,它具有小巧、高效、跨平台、开源免费和易操作的特点。

SQLite 数据库是使用 C 语言来编写的,在 CrossApp 中使用也是非常容易的。

CrossApp 已经添加了 SQLite,在 CrossApp\extensions\sqlite3 目录,按照如下步骤使用。

1. 引入头文件

```
#include "CrossAppExt.h"
```

2. 创建数据库

```
//数据库指针
sqlite3 *pdb = NULL;
//保存数据库的路径
std::string path = FileUtils::getInstance()->getWritablePath() + "save.db";
std::string sql;
int result;
//打开一个数据,如果该数据库不存在,则创建一个新的数据库文件
result = sqlite3_open(path.c_str(),&pdb);
if(result!= SQLITE_OK)
{
    CCLog("open database failed, number %d",result);
}
```

3. 执行 SQL 语句

```
//创建数据库表的 SQL 语句
sql = "create table student(ID integer primary key autoincrement,name text,sex text)";
//创建表格
result = sqlite3_exec(pdb,sql.c_str(),NULL,NULL,NULL);
if(result!= SQLITE_OK)
    CCLog("create table failed");
//向表内插入 3 条数据
sql = "insert into student values(1,'student1','male')";
result = sqlite3_exec(pdb,sql.c_str(),NULL,NULL,NULL);
if(result!= SQLITE_OK)
  CCLog("insert data failed!");
```

```cpp
sql = "insert into student values(2,'student2','female')";
result = sqlite3_exec(pdb,sql.c_str(),NULL,NULL,NULL);
if(result!= SQLITE_OK)
    CCLog("insert data failed!");
sql = "insert into student values(3,'student3','male')";
result = sqlite3_exec(pdb,sql.c_str(),NULL,NULL,NULL);
if(result!= SQLITE_OK)
    CCLog("insert data failed!");
//查询结果
char **re;
//行、列
int r,c;
//查询数据
sqlite3_get_table(pdb,"select * from student",&re,&r,&c,NULL);
CCLog("row is %d,column is %d",r,c);
//将查询出的数据通过 CCLog 输出
for(int i=1;i<=r;i++)
{
    for(int j=0;j<c;j++)
    {
        CCLog("%s",re[i*c+j]);
    }
}
sqlite3_free_table(re);
```

打印结果如图 4-2 所示。

```cpp
sql = "delete from student where ID=1";
//删除 id=1 的学生的信息
result = sqlite3_exec(pdb,sql.c_str(), NULL,NULL,NULL);
if(result!= SQLITE_OK)
    CCLog("delete data failed!");
```

图 4-2 使用 SQLite 打印结果

4. 关闭数据库

使用 SQLite 时一定要注意内存管理问题，即打开数据库，数据操作完成之后，一定要关闭数据库，否则会造成内存泄漏。关闭数据库语句如下：

```
sqlite3_close(pdb);
```

5. SQLite 保存路径

（1）Android 平台数据库保存路径如下：

```
/data/data/com.youCompany.Helloworld/files/save.db
```

（2）iOS 平台数据保存路径如下：

```
位于程序沙盒的文档目录下
../Documents/save.db
```

4.3 JSON 解析

CrossApp 使用 lib_json 来解析 JSON 文件。lib_json 已经加入了 libExtensions 下，在 CrossApp 中使用非常的便捷。

下面介绍几个类名和函数。

（1）Value

写过脚本和弱语言的读者应该很清楚 var，其实 Value 和 var 类似，都是可以表示很多数据类型的数据类型。即 Value 可以是 int，也可以是 string，或是其他数据类型。

当然 Value value 语句只是个定义，还没有决定其数据类型，如果定义语句为 Value value = 10，那么 value 的数据类型是整型。

用于 JSON 时，常表示为一个 map，其中包括 key-value，即键值对。其中 Value 包括一些将其转换为基础数据类型的如下 6 个方法：

```
value.asCString();
value.asString();
value.asBool();
value.asDouble();
value.asInt();
value.asUInt();
```

（2）FastWriter

常用函数为 write(<#const Json::Value &root#>)。

作用是将 Value 数据编码成 JSON 格式的数据。

（3）Json::Reader reader：

常用函数为 reader.parse(<#std::istream &is#>, <#Json::Value &root#>)。

作用是将 JSON 格式的数据解析为 Value 数据类型。

1. 引入头文件

若想使用 JSON 解析首先要包含 CrossAppExt.h，如下：

```
#include "CrossAppExt.h"
using namespace CSJson;
```

2. JSON 数据生成

```
//先定义数据
Value map;
map["name"] = "9miao";
map["password"] = "123456";
map["Email"] = "9miao@longtugame.com";
map["PHONE"] = 10086;
//编码成 JSON 数据
FastWriter  write;
string jsonData = write.write(map);
//打印结果
CCLog("jsonData:%s", jsonData.c_str());
```

上面的打印结果为

```
jsonData:{"Email":"9miao@longtugame.com","PHONE":10086,"name":"9miao","password":"123456"}
```

3. JSON 数据解析

有时候需要解析 Resources 目录下的 JSON 文件，首先就需要将 JSON 文件复制到 Resources 目录下。我们将下面的 JSON 格式文件复制到 Resources 目录下命名为 info.json。

```
{
    "info":
    [
        {"name":"aaa","num":"0001"},
        {"name":"bbb","num":"0002"},
        {"name":"ccc","num":"0003"},
        {"name":"ddd","num":"0004"},
        {"name":"eee","num":"0005"},
        {"name":"fff","num":"0006"},
        {"name":"ggg","num":"0007"},
        {"name":"hhh","num":"0008"},
        {"name":"iii","num":"0009"},
        {"name":"jjj","num":"0010"},
        {"name":"kkk","num":"0011"},
        {"name":"lll","num":"0012"},
        {"name":"mmm","num":"0013"},
        {"name":"nnn","num":"0014"},
```

```
        {"name":"ooo","num":"0015"},
        {"name":"ppp","num":"0016"}
    ],
    "gender": "male",
    "occupation": "coder"
}
```

在程序中添加以下代码对其解析:

```
Reader reader;
//定义 Value
Value value;
//JSON 文件路径
string jsonFile = CCFileUtils::sharedFileUtils()->fullPathForFilename("info.json");
//将文件生成 CCString 对象
CCString * jsonData = CCString::createWithContentsOfFile(jsonFile.c_str());
//将数据解析到 value 中
if (reader.parse(jsonData->getCString(),value))
{
    int length = value["info"].size();
    //循环解析子节点
    for (size_t index = 0; index < length; index++)
    {
        std::string name = value["info"][index]["name"].asString();
        std::string num = value["info"][index]["num"].asString();
        CCLog("name:%s", name.c_str());
        CCLog("num:%s", num.c_str());
    }
    //获取方式一
    Value valueGender;
    valueGender = value.get("gender", valueGender);
    std::string gender = valueGender.asCString();
    //获取方式二
    std::string occupation = value["occupation"].asCString();
    CCLog("gender:%s", gender.c_str());
    CCLog("occupation:%s", occupation.c_str());
}
```

4.4 XML 解析

CrossApp 已经加入了 tinyxml2 库用于 XML 解析。

#include "CrossApp.h"时已经包含 tinyxml2.h,因此无须再引入头文件,这样就能在开发时方便地生成和解析 XML 文件。

1. XML 文档生成

(1) 命名空间

```
using namespace tinyxml2;
```

(2) 生成 XML 代码

```cpp
//获得XML的保存路径
std::string filePath = CCFileUtils::sharedFileUtils()->getWritablePath() + "test.xml";
//生成一个XMLDocument对象
tinyxml2::XMLDocument * pDoc = new tinyxml2::XMLDocument();
//XML声明(参数可选)
XMLDeclaration * pDel = pDoc->NewDeclaration("xml version = \"1.0\" encoding = \"UTF-8\"");
//将pDel添加到XMLDocument对象中
pDoc->LinkEndChild(pDel);
//添加pTable节点
XMLElement * pTableElement = pDoc->NewElement("pTable");
//设置版本
pTableElement->SetAttribute("version", "1.0");
//将pDec添加到XMLDocument对象中
pDoc->LinkEndChild(pTableElement);
//添加一行注释
XMLComment * commentElement = pDoc->NewComment("this is xml comment");
//将注释添加到XMLDocument对象中
pTableElement->LinkEndChild(commentElement);
//添加dic节点
XMLElement * dicElement = pDoc->NewElement("dic");
pTableElement->LinkEndChild(dicElement);
//添加key节点
XMLElement * keyElement = pDoc->NewElement("key");
keyElement->LinkEndChild(pDoc->NewText("Text"));
dicElement->LinkEndChild(keyElement);
XMLElement * arrayElement = pDoc->NewElement("array");
dicElement->LinkEndChild(arrayElement);
for (int i = 0; i<3; i++) {
    XMLElement * elm = pDoc->NewElement("name");
    elm->LinkEndChild(pDoc->NewText("9miao"));
    arrayElement->LinkEndChild(elm);
}
pDoc->SaveFile(filePath.c_str());
CCLog("path:%s", filePath.c_str());
pDoc->Print();
delete pDoc;
```

(3) 生成 XML 如下：

```
<?xml version = "1.0" encoding = "UTF-8"?>
```

```xml
<pTable version = "1.0">
<!-- this is xml comment -->
<dic>
    <key>Text</key>
    <array>
        <name>9miao</name>
        <name>9miao</name>
        <name>9miao</name>
    </array>
</dic>
</pTable>
```

2. 解析 XML

下面来解析上面生成的 XML 文档。

（1）解析代码

```cpp
//解析 XML 的路径
std::string filePath = 
CCFileUtils::sharedFileUtils()->getWritablePath() + "test.xml";
    //生成一个 XMLDocument 对象
    tinyxml2::XMLDocument *pDoc = new tinyxml2::XMLDocument();
    //将 xml 文件读取到 XMLDocument 对象中
    XMLError errorId = pDoc->LoadFile(filePath.c_str());
    if (errorId != 0) {
        //xml 格式错误
        return;
    }
    XMLElement *rootEle = pDoc->RootElement();
    //获取第一个节点属性
    const XMLAttribute *attribute = rootEle->FirstAttribute();
    //打印节点属性名和值
    CCLog("attribute_name = %s,attribute_value = %s",
attribute->Name(),attribute->Value());
    XMLElement *dicEle = rootEle->FirstChildElement("dic");
    XMLElement *keyEle = dicEle->FirstChildElement("key");
    if (keyEle) {
        CCLog("keyEle Text = %s",keyEle->GetText());
    }
    XMLElement *arrayEle = keyEle->NextSiblingElement();
    XMLElement *childEle = arrayEle->FirstChildElement();
    while (childEle) {
        CCLog("childEle Text = %s",childEle->GetText());
        childEle = childEle->NextSiblingElement();
    }
    delete pDoc;
```

(2) 打印结果

```
attribute_name = version,attribute_value = 1.0
keyEle Text = Text
childEle Text = 9miao
childEle Text = 9miao
childEle Text = 9miao
```

注意,tinyxml 在 Android 上解析 assert 文件夹会有问题,解决方式如下:

```
std::string filePath = CCFileUtils::sharedFileUtils()->fullPathFromRelativeFile("test.xml");
```

第 5 章 CrossApp 设备功能调用

在移动开发过程中,常常要遇到调用移动设备的需求,如调用摄像头、通讯录、WiFi 列表和蓝牙等。CrossApp 提供了调用 Android 和 iOS 平台设备的统一接口,只需要在 CrossApp 统一地调用,然后在 Android 和 iOS 平台设置相对的权限,这样引擎会根据不同的平台去调用相应的执行,达到一样的效果。

CrossApp 引擎提供设备调用的类为 CADevice,它在引擎的 extensions\device 目录下。想要使用 CADevice 时,需要引入:

```
#include "CrossAppExt.h"
```

并声明命名空间:

```
USING_NS_CC_EXT;
```

表 5-1 为 CADevice 提供的常用函数。

表 5-1 CADevice 常用函数

函 数	说 明
void openCamera(CAMediaDelegate * target)	调用摄像头
void openAlbum(CAMediaDelegate * target)	调用相册
void startLocation(CALocationDelegate * target)	调用 GPS
float getScreenBrightness()	获得屏幕亮度
void setScreenBrightness(float brightness)	设置屏幕亮度
void writeToSavedPhotosAlbum(const std::string &s)	保存到相册
std::vector<CAAddressBookRecord> getAddressBook()	获得通讯录信息
void updateVersion(const std::string &url, unsigned int versionNumber, const std::string &appId)	更新版本
CANetWorkType getNetWorkType()	获得当前网络类型(3G 或 WiFi)
void getWifiList(CAWifiDelegate * target)	获得 WiFi 列表
void setVolume(float sender, int type)	设置音量
float getVolume(int type)	获得音量
void OpenURL(const std::string &url)	调用浏览器打开 URL 地址

续表

函　　数	说　　明
float getBatteryLevel()	获得电池电量
bool isNetWorkAvailble()	网络是否可用
void sendLocalNotification(const char * title,const char * content, unsigned long time)	发送本地通知
CAWifiInfo getWifiConnectionInfo()	获得 WiFi 详细信息
void initBlueTooth(CABlueToothDelegate * target)	初始化蓝牙
void setBlueToothType(CABlueToothType type)	对蓝牙操作（打开和关闭等）

通过上面的函数列表可以清楚地知道 CrossApp 提供的设备调用函数，但有时还需要实现相应的代理，才能获得想要的信息。比如调用摄像机拍照后，想要获得照片，必须要实现 CAMediaDelegate 代理。

下面介绍这些代理函数。

1. 摄像头代理

```
class CAMediaDelegate
{
public:
    virtual ~CAMediaDelegate(){};
    //获得照片纹理
    virtual void getSelectedImage(CAImage * image) = 0;
};
```

2. 蓝牙代理

```
class CABlueToothDelegate
{
public:
    virtual ~CABlueToothDelegate(){};
    //蓝牙状态
    virtual void getBlueToothState(CABlueToothState state){};
    //蓝牙设备信息
    virtual void getSearchBlueToothDevice(CABlueToothUnit unit){};
    //开始查找蓝牙设备
    virtual void startDiscoveryBlueToothDevice(){};
    //找到蓝牙设备
    virtual void finishedDiscoveryBlueToothDevice(){};
}
```

3. WiFi 代理

```
class CAWifiDelegate
{
public:
```

```
    virtual ~CAWifiDelegate(){};
    //WiFi信息
    virtual void getWifiListFunc(std::vector<CAWifiInfo> _wifiInfoList) = 0;
};
```

4. GPS 代理

```
class CALocationDelegate
{
public:
    virtual ~CALocationDelegate(){};
    //位置信息
    virtual void getLocations(CCDictionary * locations) = 0;
};
```

除了在 CrossApp 实现调用接口外,还需要在相应的平台配置对应的权限,有兴趣的读者可以去查找相应的权限设置对照表,这里不再赘述。

5.1 相机

上面已经介绍了 CADevice 中的一些函数和代理,本节我们就学习调用相机的实例代码。本节将要实现一个含有按钮的界面,当单击按钮时打开摄像头,拍照结束后把拍摄的照片设置为背景图片。代码需要在 Android 和 iOS 真机上演示才有效果。

首先我们在 FirstViewController.h 中引入头文件和声明命名空间:

```
#include "CrossAppExt.h"
USING_NS_CC_EXT;
```

然后要实现 CAMediaDelegate 代理,具体代码及说明如下。

(1) FirstViewController.h 文件

```
#ifndef __HelloCpp__ViewController__
#define __HelloCpp__ViewController__
#include <iostream>
#include "CrossApp.h"
#include "CrossAppExt.h"
USING_NS_CC;
USING_NS_CC_EXT;

class FirstViewController: public CAViewController,CAMediaDelegate
{
public:

    FirstViewController();
    virtual ~FirstViewController();
    //获得照片纹理
```

```cpp
    virtual void getSelectedImage(CAImage * image);
protected:
    void viewDidLoad();
    void viewDidUnload();
    //按钮回调
    void OpenCameraNow(CAControl * btn,CCPoint point);
private:
    CAImageView * imageView;
};
#endif /* defined(__HelloCpp__ViewController__) */
```

代码解析:

以上定义了视图控制器,并实现了 CAMediaDelegate,用来处理拍照的回调。

(2) FirstViewController.cpp 文件

```cpp
#include "FirstViewController.h"
FirstViewController::FirstViewController()
{

}
FirstViewController::~FirstViewController()
{

}
void FirstViewController::viewDidLoad()
{
    CADipSize size = this->getView()->getBounds().size;
    //设置背景图片
    imageView = CAImageView::createWithImage(CAImage::create("HelloWorld.png"));
    imageView->setFrame(this->getView()->getBounds());
    this->getView()->addSubview(imageView);
    //生成按钮调用摄像头
    CAButton * btn = CAButton::create(CAButtonTypeRoundedRect);
    btn->setCenter(CADipRect(size.width * 0.5, size.height * 0.5, 200, 100));
    btn->setTitleForState(CAControlStateAll,UTF8("打开照相机"));
    btn->addTarget(this,CAControl_selector(FirstViewController::OpenCameraNow),
        CAControlEventTouchUpInSide);
    this->getView()->addSubview(btn);
}
void FirstViewController::OpenCameraNow(CAControl * btn, CCPoint point)
{
    //绑定代理
    CADevice::openCamera(this);
}
void FirstViewController::getSelectedImage(CAImage * image)
{
    //将拍摄照片设置为背景图片
    imageView->setImage(image);
}

void FirstViewController::viewDidUnload()
```

```
{
    // Release any retained subviews of the main view.
    // e.g. self.myOutlet = nil;
}
```

代码解析：

以上代码在用户按下按钮后将会执行：

CADevice::openCamera(this);

这时将会弹出相机进行拍照，由于传递了参数 this，当拍照完成后会更新背景图片。

运行在 Android 设备上时，该程序需要在 AndroidManifest.xml 文件中配置摄像头的权限：

```
<uses-permission android:name="android.permission.CAMERA"/>
```

5.2 相册

上一节我们学习了如何调用摄像头，本节我们来学习如何调用系统相册。同样是打开相册，将相册中的一张图片设置为背景。

首先在 FirstViewController.h 中引入头文件和声明命名空间。

```
#include "CrossAppExt.h"
USING_NS_CC_EXT;
```

要实现 CAMediaDelegate 代理，具体代码和说明如下。

（1）FirstViewController.h 文件

```
#ifndef __HelloCpp__ViewController__
#define __HelloCpp__ViewController__
#include <iostream>
#include "CrossApp.h"
#include "CrossAppExt.h"
USING_NS_CC;
USING_NS_CC_EXT;
class FirstViewController : public CAViewController, CAMediaDelegate
{
public:
    FirstViewController();
    virtual ~FirstViewController();
    //获得图片纹理
    virtual void getSelectedImage(CAImage * image);
protected:
    void viewDidLoad();
    void viewDidUnload();
```

```
    //按钮回调(打开相册)
    void OpenAlbumNow(CAControl * btn, CCPoint point);
private:
    CAImageView * imageView;
};
#endif /* defined(__HelloCpp__ViewController__) */
```

代码解析:

以上定义了视图控制器并实现了 CAMediaDelegate 用来处理相册选择完成之后的回调。

(2) FirstViewController.cpp 文件

```
#include "FirstViewController.h"
FirstViewController::FirstViewController()
{
}
FirstViewController::~FirstViewController()
{
}
void FirstViewController::viewDidLoad()
{
    // Do any additional setup after loading the view from its nib.
    CADipSize size = this->getView()->getBounds().size;
    //创建背景
    imageView = CAImageView::createWithImage(CAImage::create("HelloWorld.png"));
    imageView->setFrame(this->getView()->getBounds());
    this->getView()->addSubview(imageView);
    //创建按钮
    CAButton * btn = CAButton::create(CAButtonTypeRoundedRect);
    btn->setCenter(CADipRect(size.width * 0.5, size.height * 0.5, 200, 100));
    btn->setTitleForState(CAControlStateAll, UTF8("打开相册"));
    btn->addTarget(this,CAControl_selector(FirstViewController::OpenAlbumNow),
        CAControlEventTouchUpInSide);
    this->getView()->addSubview(btn);
}

void FirstViewController::OpenAlbumNow(CAControl * btn, CCPoint point)
{
    //打开相册
    CADevice::openAlbum(this);
}
void FirstViewController::getSelectedImage(CAImage * image)
{
    //将选择照片设为背景
    imageView->setImage(image);
}
```

```cpp
void FirstViewController::viewDidUnload()
{
    // Release any retained subviews of the main view.
    // e.g. self.myOutlet = nil;
}
```

代码解析:

在用户按下按钮之后将执行以下语句打开相册:

CADevice::openAlbum(this);

用户在相册中选中图片后回调下面函数,将选中的图片设置为背景:

```cpp
void FirstViewController::getSelectedImage(CAImage * image)
{
    imageView->setImage(image);
}
```

该项目在 Android 设备上运行时还需要在 AndroidManifest.xml 文件中配置摄像头的权限如下:

`<uses-permission android:name="android.permission.CAMERA"/>`

5.3 通讯录

本节我们学习如何调用通讯录信息,并将通讯录的信息显示在 CATableView 上。

(1) FirstViewController.h 文件

```cpp
#ifndef __HelloCpp__ViewController__
#define __HelloCpp__ViewController__
#include <iostream>
#include "CrossApp.h"
#include "CrossAppExt.h"
USING_NS_CC_EXT;
//设置 CALabel 的 tag 值
typedef enum
{
    NOTICE = 200,
    FULLNAME,
    NUMBER,
    PROVINCE,
    NICKNAME
}ADDRESSBOOKLIST_TAG;
class FirstViewController : public CAViewController, public CATableViewDelegate,
                            public CATableViewDataSource
{
```

```cpp
public:
    FirstViewController();
    virtual ~FirstViewController();
protected:
    void viewDidLoad();
    void viewDidUnload();
public:
    virtual void tableViewDidSelectRowAtIndexPath(CATableView* table,
            unsigned int section, unsigned int row);
    virtual void tableViewDidDeselectRowAtIndexPath(CATableView* table,
            unsigned int section, unsigned int row);
    virtual CATableViewCell* tableCellAtIndex(CATableView* table,
            const CCSize& cellSize, unsigned int section, unsigned int row);
    virtual CAView* tableViewSectionViewForHeaderInSection(CATableView* table,
            const CCSize& viewSize, unsigned int section);
    virtual CAView* tableViewSectionViewForFooterInSection(CATableView* table,
            const CCSize& viewSize, unsigned int section);
    virtual unsigned int numberOfRowsInSection(CATableView *table, unsigned int section);
    virtual unsigned int numberOfSections(CATableView *table);
    virtual unsigned int tableViewHeightForRowAtIndexPath(CATableView* table,
            unsigned int section, unsigned int row);
    virtual unsigned int tableViewHeightForHeaderInSection(CATableView* table,
            unsigned int section);
    virtual unsigned int tableViewHeightForFooterInSection(CATableView* table,
            unsigned int section);
private:
    void loadingView(void);
    void getAddressBookList(void);
    void addressBookLoadProgress(float interval);
private:
    //获取屏幕大小
    CADipSize size;
    //全局的 CATableView
    CATableView* table;
    //用于保存通讯录列表信息
    std::vector<CAAddressBookRecord> addressBookList;
    //显示 CAImageView 的图片
    CAImageView* loadIamge;
    //读取进度
    int cout;
};
#endif /* defined(__HelloCpp__ViewController__) */
```

（2）FirstViewController.cpp 文件

```cpp
#include "FirstViewController.h"
//颜色
```

```cpp
#define CAColor_blueStyle ccc4(51,204,255,255)
//用于设置 tag 值
#define MAXZORDER 100
//设置调用的 Label 的宏
#define LABEL_PROPERTY(label,num)
    label->setTag(num);
    label->setColor(CAColor_blueStyle);
    label->setFontSize(_px(30));
    label->setTextAlignment(CATextAlignmentCenter);
    label->setVerticalTextAlignmet(CAVerticalTextAlignmentCenter);
FirstViewController::FirstViewController()
{
}
FirstViewController::~FirstViewController()
{
}
void FirstViewController::viewDidLoad()
{
    // Do any additional setup after loading the view from its nib.
    size = this->getView()->getBounds().size;
#if(CC_TARGET_PLATFORM!= CC_PLATFORM_WIN32)
    //如果不是 Windows 平台
    loadingView();
    getAddressBookList();
#endif
    //如果是 Windows 平台
    table = CATableView::createWithCenter(CADipRect(size.width * 0.5, size.height * 0.5,
             size.width, size.height));
    table->setTableViewDataSource(this);
    table->setTableViewDelegate(this);
    table->setAllowsSelection(true);
    table->setSeparatorViewHeight(1);
    this->getView()->addSubview(table);
}
void FirstViewController::loadingView(void)
{
    //本方法主要生成一个 loading,表示正在读取通讯录的状态
    //生成一个 View
    CAView* loading = CAView::createWithColor(ccc4(255, 255, 255, 0));
    //设置 tag 值为 MAXZORDER
    loading->setTag(MAXZORDER);
    loading->setFrame(CADipRect(this->getView()->getBounds()));
    this->getView()->addSubview(loading);
    //生成一个 CAImgaeView
    loadIamge = CAImageView::createWithImage(CAImage::create("loading.png"));
    loadIamge->setCenterOrigin(CADipPoint(size.width * 0.5, size.height * 0.5));
    loadIamge->setScale(0.5);
```

```cpp
        loading->addSubview(loadIamge);
        //开启定时器,调用 addressBookLoadProgress
        CAScheduler::schedule(schedule_selector(FirstViewController::addressBookLoadProgress),
            this, 0.01, false);
}
void FirstViewController::getAddressBookList(void)
{
    //获得通讯录列表信息
    addressBookList = CADevice::getAddressBook();
    //如果通讯录中数据大于 0 条
    if (addressBookList.size()>0)
    {
        //移除定时器,取消调用 addressBookLoadProgress
        CAScheduler::unschedule(
            schedule_selector(FirstViewController::addressBookLoadProgress), this);
        //一次 Loading
        this->getView()->removeSubviewByTag(MAXZORDER);
    }
}
void FirstViewController::addressBookLoadProgress(float interval)
{
    cout++;
    //设置角度
    loadIamge->setRotation(cout * 3);
}
void FirstViewController::viewDidUnload()
{
}
void FirstViewController::tableViewDidSelectRowAtIndexPath(CATableView * table, unsigned int section, unsigned int row)
{
}
void FirstViewController::tableViewDidDeselectRowAtIndexPath(CATableView * table, unsigned int section, unsigned int row)
{
}

CATableViewCell * FirstViewController::tableCellAtIndex(CATableView * table, const CCSize& cellSize, unsigned int section, unsigned int row)
{
    CADipSize _size = cellSize;
    //根据表示获得 cell
    CATableViewCell * cell = table->dequeueReusableCellWithIdentifier("CrossApp");
    if (cell == NULL)
    {
        cell = CATableViewCell::create("CrossApp");
#if(CC_TARGET_PLATFORM == CC_PLATFORM_WIN32)
```

```cpp
            //如果是Windows平台
            if (row == 3)
            {
                CALabel * notice = CALabel::createWithCenter(CADipRect(_size.width * 0.5,
                    _size.height * 0.5, _size.width, _size.height));
                notice->setText(UTF8("需要一个有通讯录的手机去测试"));
                LABEL_PROPERTY(notice, NOTICE);
                cell->addSubview(notice);
            }
#else
            //如果不是Windows平台
            //姓名
            CALabel * fullName = CALabel::createWithFrame(CADipRect(_size.width * 0.01, 0,
                            _size.width * 0.3, _size.height));
            LABEL_PROPERTY(fullName, FULLNAME);
            cell->addSubview(fullName);
            //电话号码
            CALabel * phoneNumber = CALabel::createWithFrame(CADipRect(_size.width * 0.3, 0,
                              _size.width * 0.3, _size.height));
            LABEL_PROPERTY(phoneNumber, NUMBER);
            cell->addSubview(phoneNumber);
            //职位
            CALabel * province = CALabel::createWithCenter(CADipRect(_size.width * 0.7,
                            _size.height * 0.5, _size.width * 0.2, _size.height));
            LABEL_PROPERTY(province, PROVINCE);
            cell->addSubview(province);
            //昵称
            CALabel * nickName = CALabel::createWithCenter(CADipRect(_size.width * 0.9,
                            _size.height * 0.5, _size.width * 0.2, _size.height));
            LABEL_PROPERTY(nickName, NICKNAME);
            cell->addSubview(nickName);
#endif
        }
#if(CC_TARGET_PLATFORM!= CC_PLATFORM_WIN32)
    CALabel * fullName = (CALabel *)cell->getSubviewByTag(FULLNAME);
    fullName->setText(addressBookList[row].fullname.c_str());
    CALabel * phoneNumber = (CALabel *)cell->getSubviewByTag(NUMBER);
    phoneNumber->setText(addressBookList[row].phoneNumber.c_str());
    CALabel * province = (CALabel *)cell->getSubviewByTag(PROVINCE);
    province->setText(addressBookList[row].province.c_str());
    CALabel * nickName = (CALabel *)cell->getSubviewByTag(NICKNAME);
    nickName->setText(addressBookList[row].nickname.c_str());
#endif
    return cell;
}
CAView * FirstViewController::tableViewSectionViewForHeaderInSection(CATableView * table,
    const CCSize& viewSize, unsigned int section)
```

```cpp
{
    CAView* view = CAView::createWithColor(CAColor_gray);
    return view;
}
CAView* FirstViewController::tableViewSectionViewForFooterInSection(CATableView* table,
const CCSize& viewSize, unsigned int section)
{
    CAView* view = CAView::createWithColor(CAColor_gray);
    return view;
}
unsigned int FirstViewController::numberOfRowsInSection(CATableView* table, unsigned int section)
{
#if(CC_TARGET_PLATFORM == CC_PLATFORM_WIN32)
    //如果是Windows平台则返回8
    return 8;
#endif
    //如果不是Windows平台则返回通讯录有多少条
    return addressBookList.size();
}

unsigned int FirstViewController::numberOfSections(CATableView* table)
{
    return 1;
}

unsigned int FirstViewController::tableViewHeightForRowAtIndexPath(CATableView* table,
        unsigned int section, unsigned int row)
{
#if(CC_TARGET_PLATFORM == CC_PLATFORM_WIN32)
    //如果是Windows平台高度为1/8
    return size.height / 8;
#endif
    //如果不是Windows平台高度为1/10
    return size.height * 0.1;

}
unsigned int FirstViewController::tableViewHeightForHeaderInSection(CATableView* table,
    unsigned int section)
{
    return 1;
}
unsigned int FirstViewController::tableViewHeightForFooterInSection(CATableView* table,
unsigned int section)
{
    return 1;
}
```

代码解析：

该案例在 Android 平台运行需要在 AndroidManifest.xml 添加下面的权限：

```
<!-- 读取联系人权限 -->
<uses-permission android:name="android.permission.READ_CONTACTS"/>
```

5.4　本章小结

CrossApp 提供的 WiFi、蓝牙和音量等接口都非常的类似，本节就不再赘述。CrossApp 还在不断的开发当中，其中有些设备调用的功能正在研发中，功能并不稳定，调用设备时遇到崩溃的 bug 请及时去 CrossApp 的官方论坛上反馈解决。

第 6 章 CrossApp 多媒体

本章我们将学习 CrossApp 的多媒体功能,即如何设置动画,播放音效和播放视频等。

6.1 CAViewAnimation 动画

CrossApp 引擎为我们提供动画效果的是 CAViewAnimation,有过 iOS 开发经验的读者使用起来比较熟悉,因为其基本构成是参照动画来设计的。

CAViewAnimation 的常用函数如表 6-1 所示,都为 static,可以直接调用。

表 6-1 CAViewAnimation 常用函数

函 数	说 明
static void beginAnimations(const std::string& animationID, void * context);	开始执行动画
static void setAnimationDuration(float duration)	动画播放时长(默认 0.2s)
static void setAnimationDelay(float delay)	动画延时播放时长(默认 0s)
static void setAnimationCurve(const CAViewAnimationCurve& curve)	动画波频控制 CAViewAnimationCurveLinear CAViewAnimationCurveEaseOut CAViewAnimationCurveEaseIn CAViewAnimationCurveEaseInOut
static void setAnimationWillStartSelector(CAObject * target, SEL_CAViewAnimation0 selector);	动画开始前的回调(零参数)
static void setAnimationWillStartSelector(CAObject * target, SEL_CAViewAnimation2 selector);	动画开始前的回调(两参数)
static void setAnimationDidStopSelector(CAObject * target, SEL_CAViewAnimation0 selector)	动画结束后的回调(零参数)
static void setAnimationDidStopSelector(CAObject * target, SEL_CAViewAnimation2 selector)	动画结束后的回调(两参数)
static void commitAnimations()	执行动画
static bool areAnimationsEnabled()	动画是否执行完毕
static bool areBeginAnimations()	动画是否执行完成

接下来介绍如何使用 CAViewAnimation 来创建一个动画，实例代码如下

```
void FirstViewController::viewDidLoad()
{
    CCRect winRect = this->getView()->getBounds();
    CAImageView* imageView =
        CAImageView::createWithImage(CAImage::create("HelloWorld.png"));
    imageView->setFrame(winRect);
    this->getView()->addSubview(imageView);
    CALabel* label = CALabel::createWithCenter(CCRect(winRect.size.width * 0.5,
        winRect.size.height * 0.5 - 270, winRect.size.width, 200));
    label->setTextAlignment(CATextAlignmentCenter);
    label->setVerticalTextAlignmet(CAVerticalTextAlignmentCenter);
    label->setFontSize(72 * CROSSAPP_ADPTATION_RATIO);
    label->setText("Hello World!");
    label->setColor(CAColor_white);
    this->getView()->insertSubview(label, 1);
    //开始执行动画
    CAViewAnimation::beginAnimations("Rotation", NULL);
    //动画时长
    CAViewAnimation::setAnimationDuration(1.0f);
    //动画延迟时长执行
    //CAViewAnimation::setAnimationDelay(0.3f);
    //动画波频控制
    CAViewAnimation::setAnimationCurve(CAViewAnimationCurveLinear);
    //动画开始前回调(两参数)
    CAViewAnimation::setAnimationWillStartSelector(this,
        CAViewAnimation2_selector(FirstViewController::willStartAction));
    //动画完成回调(两参数)
    CAViewAnimation::setAnimationDidStopSelector(this,
        CAViewAnimation2_selector(FirstViewController::didStopAction));
    //设置 CAView 的变化,必须在 beginAnimations 和 commitAnimations 之间
    label->setColor(CAColor_yellow);
    //旋转
    label->setRotation(-1080);
    label->setFrame(CADipRect(winRect.size.width * 0.1, 200, 150, 150));
    //执行动画
    CAViewAnimation::commitAnimations();
}
void FirstViewController::willStartAction(const string& animationID, void* context)
{
    //动画开始回调
```

```cpp
    CCLog("willStartAction");
}

void FirstViewController::didStopAction(const string& animationID, void * context)
{
    //动画结束回调
    CCLog("didStopAction");
}
```

这样我们就设置了一个改变 CALabel 的颜色和旋转及改变 Frame 的动画，这里要注意的是所有设置 CAView 动画变化的数值必须写在 beginAnimations 和 commitAnimations 之间，否则不会执行动画，而是立即改变了 CAView 的数值。

6.2 SimpleAudioEngine 音效

CrossApp 提供的音频库位于 CocosDenshion 中，其接口由 SimpleAudioEngine 定义，提供了基本的背景音乐和音效播放。

SimpleAudioEngine 的实现是跨平台的，在 Windows 平台上由 MCI 相关 API 实现；在 Android 平台上通过 JNI 调用 Android SDK 中的 AudioPlayer 实现；而在 iOS 平台上由 Cocoa SDK 里的 Core-Audio 实现。

SimpleAudioEngine 提供的声音播放分为两种：背景音乐和音效。同一时间只能播放一首背景音乐，但是能同时播放许多音效。背景音乐支持的格式如表 6-2 所示。

表 6-2 背景音乐格式和平台

平台	支持的 BGM 格式
Android	在 Android 上 CocosDenshion 能支持的音频格式对应于 android.media.MediaPlayer 所支持的格式
iOS	在 iOS 上 Cocos2d-x 的 CocosDenshion 所支持的格式和 Cocos2d-iPhone 中所支持的一样，mp3，caf 是推荐的格式
Windows	mid 和 wav 是支持的格式，mp3 不支持

音效格式支持与平台见表 6-3。

表 6-3 音效格式和平台

平台	支持的音效格式
Android	ogg 是最好的选择，对 wav 的支持不是很好
iOS	iOS 和 Cocos2d-iPhone 中 CocosDenshion 所支持的格式一样，个人建议是苹果的 caf 格式
Windows	mid，wav

SimpleAudioEngine 被设计成了一个单例,方便全局调用。表 6-4 示出了 SimpleAudioEngine 的常用函数供我们对音乐进行操作。

表 6-4 SimpleAudioEngine 常用函数

函 数	说 明
static SimpleAudioEngine * sharedEngine()	获得 SimpleAudioEngine 对象
void preloadBackgroundMusic(const char * pszFilePath);	预加载背景音乐
void playBackgroundMusic(const char * pszFilePath, bool bLoop)	播放背景音乐,bool 值表示是否循环播放
void stopBackgroundMusic(bool bReleaseData)	停止播放背景音乐
void pauseBackgroundMusic()	暂停播放背景音乐
void resumeBackgroundMusic()	恢复播放背景音乐
void rewindBackgroundMusic()	从头播放背景音乐
bool isBackgroundMusicPlaying()	背景音乐是否在播放
float getBackgroundMusicVolume()	获得背景音乐音量
void setBackgroundMusicVolume(float volume)	设置背景音乐音量
void preloadEffect(const char * pszFilePath);	预加载音效
void unloadEffect(const char * pszFilePath)	释放音效文件
unsigned int playEffect(const char * pszFilePath, bool bLoop)	播放音效,bool 值表示是否重复播放。返回值为 soundId
void pauseEffect(unsigned int nSoundId)	根据 soundId 暂停音效
void pauseAllEffects()	暂停全部音效
void resumeEffect(unsigned int nSoundId)	根据 soundId 恢复播放音效
void resumeAllEffects()	恢复播放全部音效
void stopEffect(unsigned int nSoundId)	根据 soundId 停止播放音效
void stopAllEffects()	停止播放所有音效
float getEffectsVolume()	获得音效的音量大小
void setEffectsVolume(float volume)	设置音效的音量大小

如果想要播放背景音乐和音效,首先需要引入头文件和声明命名空间。

```
#include "SimpleAudioEngine.h"
using namespace CocosDenshion;
```

然后在需要播放背景音乐或音效的地方添加相应的逻辑。
实例代码如下:

```
//获得 SimpleAudioEngine 对象
SimpleAudioEngine * audio = SimpleAudioEngine::sharedEngine();
//预加载背景音乐
audio->preloadBackgroundMusic("bg.mp3");
//播放背景音乐,并设置循环播放
audio->playBackgroundMusic("bg.mp3", true);
```

```cpp
//暂停播放背景音乐
audio->pauseBackgroundMusic();
//恢复播放背景音乐
audio->resumeBackgroundMusic();
//停止播放背景音乐
audio->stopBackgroundMusic();
//预加载音效
audio->preloadEffect("effect.mp3");
//播放音效
unsigned int soundId = audio->playEffect("effect.mp3");
//暂停音效
audio->pauseEffect(soundId);
//恢复播放音效
audio->resumeEffect(soundId);
//停止播放全部音效
audio->stopAllEffects();
//释放音效
audio->unloadEffect("effect.mp3")
```

第 7 章 CrossApp 网络通信

本章将介绍 CrossApp 的网络通信模块，在进行 App 开发时，大量数据都要存放在服务器，无论是做新闻展示、聊天通信还是资料筛选，都希望能及时地从服务器获得最新的信息。

7.1 HTTP 基础使用

CrossApp 为我们封装了 HTTP 的网络框架，其文件在 CrossApp 引擎包的\extensions\network 目录下的 HttpClient、HttpRequest 和 HttpResponse。

进行一次 HTTP 交互，涉及如下三个类：

（1）HttpRequest 用来描述一个请求。

（2）HttpResponse 用来描述对应请求的响应。

（3）HttpClient 是一个单例模式的类，其职责就是将收到的 HttpRequest 对象 push 到发送队列中，并发送一个信号量驱动工作线程工作，工作线程再将收到的数据封装成一个 HttpResponse 对象 push 接收队列，并启用调度来派送数据。

引用头文件：

```
#include "CrossAppExt.h"
```

1. 发送 HTTP 请求

//创建 request 对象，这里新建的对象不能使用 autorelease()，原因将在后面介绍。

```
CCHttpRequest * request = newCCHttpRequest();
//设置 URL
request->setUrl("www.9miao.com");
//设置请求类型 kHttpGet、kHttpPost、KHttpPostFile、kHttpPut、kHttpDelete、kHttpUnkown
request->setRequestType(CCHttpRequest::kHttpGet);
//设置回调对象和回调函数
request->setResponseCallback(this,
        httpresponse_selector(FirstViewController::requestresult));
//设置用户标识，可以通过 response 获取
request->setTag("get");
//使用 CCHttpClient 共享实例来发送 request
```

```cpp
CCHttpClient::getInstance()->send(request);
//调用 release()
request->release();
```

2. 接收请求结果

接收请求结果的实例代码如下：

```cpp
void FirstViewController::requestresult(CCHttpClient * clinet, CCHttpResponse * response)
{
    if (!response->isSucceed())
    {
        return;
    }
    //获取返回代码,例如 200 和 404 等
    int statusCode = response->getResponseCode();
    if (!strcmp("get", response->getHttpRequest()->getTag()) && (statusCode == 200))
    {
        std::string responseRes = "";
        std::vector<char> * buffer = response->getResponseData();
        for (unsigned int i = 0; i < buffer->size(); i++)
        {
            responseRes += (*buffer)[i];
        }
        //查找字符"官方特约"
        string::size_type idx = responseRes.find(UTF8("官方特约"));
        if (idx ==-1)
        {
            //未找到字符
            CCLog(" Not Found");
            return;
        }
        string temp = responseRes.substr(idx, 30);
        string num = temp.substr(temp.find(UTF8(":")) + 3,
                    temp.find("<") - temp.find(UTF8(":")) - 3);
        CCLog("temp:%s",temp.c_str());
    }
    else
    {
        //打印返回代码
        CCLog("statusCode:%d", statusCode);
    }
}
```

7.2 HTTP 加载网络图片

本节将介绍网络图片的加载,并将其显示在本机的屏幕上。

1. 发送图片请求

```
/**
编译到 Android 平台之后注意添加联网权限
**/
        CCHttpRequest * request = new CCHttpRequest();
        //请求图片的 URL
        request->setUrl(
            "http://www.9miao.com/template/dean_hotspot_141011/deancss/logo.png");
        //请求方式
        request->setRequestType(CCHttpRequest::kHttpGet);
        //设置回调
        request->setResponseCallback(this,
            httpresponse_selector(FirstViewController::onHttpRequestImageCompleted));
        //设置 tag
        request->setTag("image");
        CCHttpClient::getInstance()->send(request);
        request->release();
```

2. 接收图片的代码

```
void FirstViewController::onHttpRequestImageCompleted(CCHttpClient * sender,
    CCHttpResponse * response)
{
        if (!response)
        {
                return;
        }
        if (!response->isSucceed())
        {
                CCLog("response failed");
                CCLog("error buffer: %s", response->getErrorBuffer());
                return;
        }
        vector<char> * buffer = response->getResponseData();
        //创建一个 CCImage
        CCImage * img = new CCImage();
        img->initWithImageData((unsigned char *)buffer->data(), buffer->size());
        //创建一个纹理 CAImage
        CAImage * texture = new CAImage();
        bool isImg = texture->initWithImage(img);
        img->release();
```

```
            //通过CAImageView绘制到屏幕
            CAImageView * iv = CAImageView::createWithImage(texture);
            iv->setFrame(CCRect(0, 0, 200, 200));
            this->getView()->addSubview(iv);
    }
```

这样我们就可以将网络的图片显示到本机的屏幕上了。

第 8 章 CrossApp 项目实战

本章将通过一款 9 秒社区开源项目折 800 来了解如何通过 CrossApp 搭建完整的跨平台 App 项目，该项目可以在 crossapp.9miao.com 下载源码。

8.1 折 800 开源项目介绍

首先下载折 800 源码，然后解压缩到 CrossApp 的安装目录下的 projects 目录下，如图 8-1 所示。

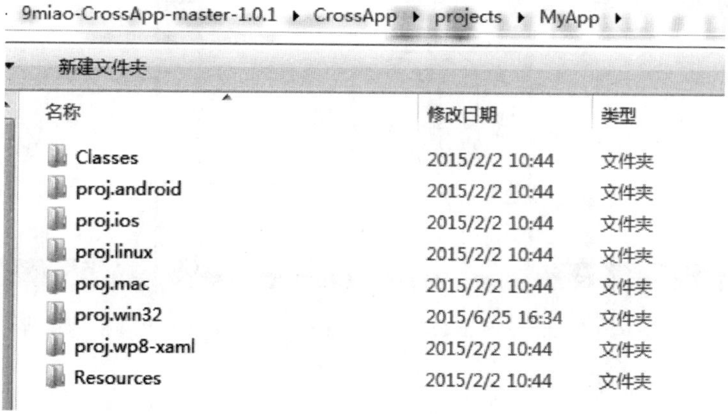

图 8-1 折 800 项目目录

之后进入 proj.win32 目录，打开 MyApp.sln，使用 Visual Studio 2013 打开项目，如图 8-2 所示。

然后在 Visual Studio 2013 中编译运行 MyApp 项目，效果如图 8-3 所示。

在调试该项目时要注意 CrossApp 版本。

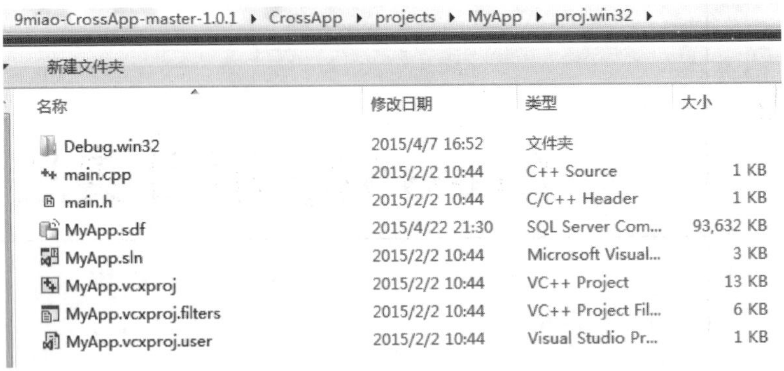

图 8-2　折 800 项目文件目录

图 8-3　CrossApp 实现的折 800 项目在 Visual Studio 2013 中的运行

8.2　项目架构设计

该项目的总体功能主要有界面的控制器切换，数据显示和动态加载，实际上是一款跨平台电商 App 的架构，代码结构如图 8-4 所示。

其中源码目录文件及功能如下：

（1）DB 目录

DB.h：数据处理的单例模式封装类

DB.cpp

DBBase.h：数据处理的基础类

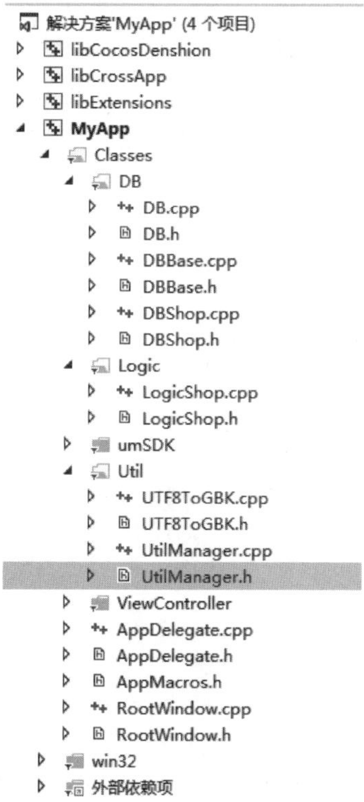

```
解决方案'MyApp' (4 个项目)
  ▷ libCocosDenshion
  ▷ libCrossApp
  ▷ libExtensions
  ▲ MyApp
    ▲ Classes
      ▲ DB
        ▷ ++ DB.cpp
        ▷    DB.h
        ▷ ++ DBBase.cpp
        ▷    DBBase.h
        ▷ ++ DBShop.cpp
        ▷    DBShop.h
      ▲ Logic
        ▷ ++ LogicShop.cpp
        ▷    LogicShop.h
      ▷ umSDK
      ▲ Util
        ▷ ++ UTF8ToGBK.cpp
        ▷    UTF8ToGBK.h
        ▷ ++ UtilManager.cpp
        ▷    UtilManager.h
      ▷ ViewController
      ▷ ++ AppDelegate.cpp
      ▷    AppDelegate.h
      ▷    AppMacros.h
      ▷ ++ RootWindow.cpp
      ▷    RootWindow.h
  ▷ win32
  ▷ 外部依赖项
```

图 8-4 折 800 项目代码结构

DBBash.cpp

DBShop.h：商店数据处理类

DBShop.cpp

（2）Logic 目录

LogicShop.h：业务逻辑的实现类

LogicShop.cpp

（3）umSDK：第三方统计平台相关类

（4）Util 目录

UTF8ToGBK.cpp

UTF8ToGBK.h：字符串转码功能类

UtilManager.cpp

UtilManager.h：对 UI 处理的功能类封装

（5）ViewConller：包含了所有 UI 界面的控制器类

（6）AppDelegate.cpp

（7）AppDelegate.h：项目入口类

（8）RootWindow.cpp

(9) RootWindow.h：根窗口类

通过以上的源码说明，可以看到，该项目主要分为了模型部分、视图部分和控制器部分，是按照 MVC 方式进行的设计。视图部分包含了各个可见窗口的封装；模型部分定义了底层的数据处理；控制器部分负责调用模型实现数据处理以及根据数据变化切换 UI 的显示。下一节将对核心模块的代码进行说明。

8.3 核心模块说明

本项目虽然没有完全商业化，但核心功能为商业化项目提供了一个很好的原型框架，首先介绍模型部分。

1. 业务模型封装

该部分通过 DB.h 和 DB.cpp 实现了一个数据处理的单例定义，DB.h 代码如下：

```cpp
#include "CrossApp.h"
#include "DBShop.h"
class DB : public CAObject{
public:
    DB();
    virtual ~DB();
    virtual bool init() { return true; }
    CC_SYNTHESIZE_RETAIN(DBShopList*, m_shopDB, ShopDB);
    static DB* shareDB();
    CREATE_FUNC(DB);
};
```

代码解析：

以上定义了一个静态函数 shareDB 来获取 DB 的单例，CC_SYNTHESIZE_RETAIN 宏定义实现了一个成员属性和 set/get 方法的自动声明。下面是 DB.cpp 代码：

```cpp
#include "DB.h"
#include "Util/UTF8ToGBK.h"
static DB* db = NULL;
DB::DB()
:m_shopDB(NULL)
{
    setShopDB(DBShopList::create());
}
DB::~DB() {
    CC_SAFE_RELEASE_NULL(m_shopDB);
    CC_SAFE_RELEASE_NULL(db);
}
DB* DB::shareDB() {
    if (db == NULL) {
```

```
            db = DB::create();
            db->retain();
            return db;
        }
        return db;
    }
```

代码解析：

以上代码通过单例模式封装了一个 DB 对象，在该对象中定义了一个成员 ShopDB，该成员描述了一组商品。DBShopList 继承 DBBase，DBBase 为一组数据集合的抽象封装，以下为 DBBase.h 源代码：

```
#include "CrossApp.h"
#include <iostream>
#include <vector>
USING_NS_CC;
using namespace std;
class DBBase : public CAObject {
public:
    DBBase();
    virtual ~DBBase();
    virtual bool init() { return true; }
    CC_SYNTHESIZE(int, m_id, ID);
    void add(CAObject* obj);
    DBBase* find(int id);
    bool remove(int id);
    bool remove(CAObject* obj);
    void removeAllObject();
    vector<int> getIDs();
    int count();
    CREATE_FUNC(DBBase);
private:
    CCArray* m_array;
    CCDictionary* m_dict;
    DBBase* m_findCache;
};
```

代码解析：

以上代码定义了一个集合 vector 以及对这个集合的管理，DBBase.cpp 详见源码，由于代码较多，此处不再赘述。

以下是 DBShopt.h 源码：

```
#include "CrossApp.h"
#include <iostream>
#include "DBBase.h"
USING_NS_CC;
```

```cpp
using namespace std;
class DBShop : public DBBase {
public:
    DBShop();
    virtual ~DBShop();
    CC_SYNTHESIZE(int, m_ID, ID);                         //商品ID
    CC_SYNTHESIZE(string, m_name, Name);                  //商品名称
    CC_SYNTHESIZE(float, m_price, Price);                 //商品价格,折扣后
    CC_SYNTHESIZE(float, m_priceOld, PriceOld);           //商品价格,折扣前
    CC_SYNTHESIZE(string, m_icon, Icon);                  //商品图标
    CC_SYNTHESIZE(int, m_number, Number);                 //商品数量
    CC_SYNTHESIZE(float, m_discount, Discount);           //商品折扣
    CREATE_FUNC(DBShop);
};
class DBShopList : public DBBase {
public:
    DBShopList();
    virtual ~DBShopList();
    CREATE_FUNC(DBShopList);
};
```

代码解析：

DBShop 继承了 DBBase，实现了商品的数据结构，在此基础上可以拓展出各种不同的的数据来继承 DBBase，比如用户数据和订单数据等。

2. 业务逻辑处理封装

在 LogicShop.h 和 LogicShop.cpp 中实现了对商店数据处理的封装，这里介绍本地数据实现的数据模拟，用户可以在此基础上扩展为网络通信，同时如果在项目中有多种数据，可以将不同类型数据的处理分别对应一个业务逻辑类，LogicShop.h 代码如下：

```cpp
#include "CrossApp.h"
#include <iostream>
#include "DB/DB.h"
USING_NS_CC;
using namespace std;
class LogicShop {
public:
    LogicShop();
    virtual ~LogicShop();
    static int getShopItemCount();
    static vector<int> getShopIDs();
    static string getShopName(int id);
    static string getShopIcon(int id);
    static float getShopPrice(int id);
    static float getShopPriceOld(int id);
    static float getShopDicount(int id);
```

```cpp
    static int getShopNumber(int id);
    static void createShopItemsIndex(int index);
    static void createShopItemsForTest();
    static void createShopItemsForPage1();
    static void createShopItemsForPage2();
    static void createShopItemsForPage3();
};
```

3. 视图和视图控制器

项目首先由一个 RootWindow 来实现主视图的加载，代码如下：

```cpp
bool RootWindow::init()
{
    if (!CAWindow::init())
    {
        return false;
    }
    startUMSDK();
    /* CAUserDefault::sharedUserDefault()->setBoolForKey("isGuide", true);
    CAUserDefault::sharedUserDefault()->flush(); */
    bool isGuide = CAUserDefault::sharedUserDefault()->getBoolForKey("isGuide",
        true);
    if (isGuide) {
        GuideViewController * guideView = new GuideViewController();
        guideView->init();
        this->setRootViewController(guideView);
        guideView->release();
    } else {
        CATabBarController * tabBar = new CATabBarController();
        tabBar->initWithViewControllers(loadControlViews());
        tabBar->setTabBarTitleColorForNormal(CAColor_gray);
        tabBar->setTabBarTitleColorForSelected(CAColor_red);
    tabBar->setTabBarBackGroundImage(
        CAImage::create("image/tab_bar_back.png"));
    tabBar->setTabBarSelectedBackGroundImage(
        CAImage::create("image/tab_bar_back.png"));
        tabBar->setTabBarSelectedIndicatorImage(NULL);
        //this->setRootViewController(tabBar);
        //tabBar->release();
        CANavigationController * rootNav = new CANavigationController();
        rootNav->setNavigationBarHidden(true, false);          //隐藏根 nav 视图导航
        rootNav->initWithRootViewController(tabBar);
        tabBar->release();
        this->setRootViewController(rootNav);
        rootNav->release();
    }
    return true;
}
```

代码解析：

在 RootWindow.cpp 中定义了导航控制器，包含了 tabBar。下面介绍导航控制器的初始化，代码如下：

```cpp
vector<CAViewController*> RootWindow::loadControlViews() {
    //firstView
    FirstViewController* first = new FirstViewController();
    first->init();
    //first->setNavigationBarItem(CANavigationBarItem::create(
    //    UTF8ToGBK::transferToGbk("折800").c_str()));
    //CANavigationBarItem* item = new CANavigationBarItem();
    //item->setTitleViewImage(
    //    CAImage::create("image/app_brand_logo_defalut.png"));
    //first->setNavigationBarItem(item);
    //item->release();
    CANavigationController* navigationController1 = new CANavigationController();
    navigationController1->setNavigationBarBackGroundImage(
        CAImage::create("image/titleBarBack.png"));
    navigationController1->initWithRootViewController(first);
    navigationController1->setTabBarItem(
        CATabBarItem::create(UTF8ToGBK::transferToGbk("首页").c_str(),
        CAImage::create("image/tag_jinri_normal.png"), CAImage::create("image/tag_jinri_process.png")));
    first->release();
    //secondView
    SecondViewControler* second = new SecondViewControler();
    second->init();
    second->setNavigationBarItem(CANavigationBarItem::create(
        UTF8ToGBK::transferToGbk("分类").c_str()));
    CANavigationController* navigationController2 = new CANavigationController();
    navigationController2->setNavigationBarBackGroundImage(
        CAImage::create("image/titleBarBack.png"));
    navigationController2->initWithRootViewController(second);
    navigationController2->setTabBarItem(
        CATabBarItem::create(UTF8ToGBK::transferToGbk("分类").c_str(),
        CAImage::create("image/tag_category_nor.png"),
        CAImage::create("image/tag_category_process.png")));
    second->release();
    //thirdView
    ThirdViewControler* third = new ThirdViewControler();
    third->init();
    third->setNavigationBarItem(CANavigationBarItem::create(
    UTF8ToGBK::transferToGbk("品牌团").c_str()));
    CANavigationController* navigationController3 = new CANavigationController();
    navigationController3->setNavigationBarBackGroundImage(
        CAImage::create("image/titleBarBack.png"));
```

```cpp
        navigationController3->initWithRootViewController(third);
        navigationController3->setTabBarItem(
            CATabBarItem::create(UTF8ToGBK::transferToGbk("品牌团").c_str(),
            CAImage::create("image/tag_brand_group_nor.png"),
            CAImage::create("image/tag_brand_group_process.png")));
    third->release();
    //fourthView
    FourthViewController * fourth = new FourthViewController();
    fourth->init();
    fourth->setNavigationBarItem(CANavigationBarItem::create(
            UTF8ToGBK::transferToGbk("积分商城").c_str()));
    CANavigationController * navigationController4 = new CANavigationController();
    navigationController4->setNavigationBarBackGroundImage(
            CAImage::create("image/titleBarBack.png"));
    navigationController4->initWithRootViewController(fourth);
    navigationController4->setTabBarItem(CATabBarItem::create(
            UTF8ToGBK::transferToGbk("积分商城").c_str(),
            CAImage::create("image/tag_back_integration_nor.png"),
            CAImage::create("image/tag_back_integration_process.png")));
    fourth->release();
    //fifthView
    FifthViewController * fifth = new FifthViewController();
    fifth->init();
    fifth->setNavigationBarItem(CANavigationBarItem::create(
            UTF8ToGBK::transferToGbk("个人中心").c_str()));
    CANavigationController * navigationController5 = new CANavigationController();
    navigationController5->setNavigationBarBackGroundImage(
            CAImage::create("image/titleBarBack.png"));
    navigationController5->initWithRootViewController(fifth);
    navigationController5->setTabBarItem(
            CATabBarItem::create(UTF8ToGBK::transferToGbk("个人中心").c_str(),
            CAImage::create("image/tag_user_center_nor.png"),
            CAImage::create("image/tag_user_center_process.png")));
    fifth->release();
    std::vector<CAViewController *> controllerItems;
    controllerItems.push_back(navigationController1);
    controllerItems.push_back(navigationController2);
    controllerItems.push_back(navigationController3);
    controllerItems.push_back(navigationController4);
    controllerItems.push_back(navigationController5);
    return controllerItems;
}
```

代码解析：

通过 RootWindow 实现了整个应用的主窗口，并且可以在底部按钮切换不同的视图控制器。关于不同的子界面实现，有兴趣的读者可以自行阅读项目源码，来了解视图和数据模型层的协作，主窗口启动以后效果如图 8-5 所示。

图 8-5　折 800 项目主界面

8.4　本章小结

本章主要介绍了一款由 9 秒社区开发的 CrossApp 开源项目，并分析了该项目的核心架构，读者可以在此基础上进行完善。